HOW TO READ

Historical Mathematics

HOW TO READ

Historical Mathematics

Benjamin Wardhaugh

PRINCETON UNIVERSITY PRESS

Princeton and Oxford

Published by Princeton University Press,
41 William Street, Princeton, New Jersey 08540

In the United Kingdom:
Princeton University Press,
6 Oxford Street,
Woodstock, Oxfordshire OX20 1TW

Library of Congress Cataloging-in-Publication Data

Wardhaugh, Benjamin, 1979–
How to read historical mathematics / Benjamin Wardhaugh.
p. cm.
Includes bibliographical references and index.
ISBN 978-0-691-14014-8 (hardcover : alk. paper)
1. Mathematics—History. 2. Guided reading. I. Title.
QA21.W324 2010
510.9—dc22   2009041704

British Library Cataloging-in-Publication Data is available

*This book has been composed in Adobe Garamond Pro
with IM Fell Double Pica Pro display*

Printed on acid-free paper. ∞

press.princeton.edu

Printed in the United States of America

7 9 10 8 6 4 2

CONTENTS

PREFACE

Read Euler, read Euler, he is the master of us all.

—Pierre-Simon Laplace, quoted by Gugliemo Libri
in the *Journal des Savants*, January 1846, p. 51

If you've ever read a historical mathematical text and then
thought, *what on earth was that?* this book is for you.

Let me explain.

Reading historical mathematics is fascinating, challenging,
enriching, and endlessly rewarding. A huge wealth of mathematics and mathematical experience are contained in the past;
the enthralling task of finding out about them can transform
your study and enjoyment of modern mathematics, and it can
turn into the study of a lifetime in its own right.

Nations and societies commemorate people and events from
their past, and by doing so they create and strengthen their own
sense of identity. Mathematicians do the same thing when they
commemorate the past in the names they use: the Isaac Newton Institute, Lebesgue integrals, Noether's theorem. . . . We
all know who Isaac Newton was, but most of us are a bit more
shaky about Henri Lebesgue or Emmy Noether. That's where
the history of mathematics comes in. Who were the people
who made our mathematics the way it is? What did they
do? Why do we remember them? (There are more awkward

questions, too: What about the people we don't remember? Why do some mathematicians and their writings become part of the mathematical "hall of fame" and others not? Why, in the end, is our mathematics the way it is?)

Answering those questions is fun, and it need not be hard. More and more people are now experiencing the fascination of the history of mathematics, through television programs, popular books, and public lectures. Courses on the history of mathematics are often offered as part of a degree in mathematics.

With those courses come sourcebooks, like the ever-popular *Reader* edited by John Fauvel and Jeremy Gray and the recent *Mathematics Emerging* edited by Jacqueline Stedall. On the web there are vast treasure troves of resources for the history of mathematics. Today you can read not just Euler but a whole range of mathematical "masters" if you want to.

Reading the mathematical "classics" is a way to enrich ourselves, to engage with our predecessors and learn from them not just what they did but also how they did it, and why. Just as reading the classics of literature can help us to see our own world with new eyes, and reading the classics of science writing can both inspire modern science and help us to remember why we are doing it, so reading historical mathematics can help us to see modern mathematics afresh, to reconnect with the distinctively mathematical way of thinking that runs through human history and to recall the reasons for doing mathematics in the first place. By reading their works, we are privileged to have access to the great minds of the past; and we learn something about who we are, too.

At a more mundane level, reading mathematics from the past can help us to learn mathematics itself: it often provides a different perspective on familiar ideas, showing what really motivated their development. And it can make mathematics more engaging by bringing to life some of the human stories which make up what is fundamentally a very human enterprise.

So, as I said above, reading historical mathematics is fascinating, enriching, and rewarding. Learning to savor the mathematical language of the past, to understand it more deeply and to enter into its authors' ways of thought, and to be inspired by it in our own intellectual endeavors are valuable and worthwhile activities.

Reading historical mathematics can also be hard. The sourcebooks I mentioned above give a lot of guidance to their readers, but it can still be daunting to approach mathematical writings from the past for the first time—particularly if you are taking a course which asks you to engage with them, think about them, or write about them.

Many things might be unfamiliar: the language, the notation, the fact that you are reading a translation rather than the author's actual words. The sources for historical mathematics were written in times and places that are now mostly unfamiliar to us, by people whose ideas and values were very different from our own and whose mathematical culture, methods, and assumptions may have been very different from anything we are familiar with. When you think about a piece of historical writing more deeply, there are still more questions: What was the actual book like that this text appeared in? Who read it? Who was meant to read it? Why are we, now, studying it? Who says that it's important, and what does that mean?

It can be hard even to know what questions to ask, let alone to find any answers. What can we really know about a piece of mathematical writing from the past, given all the distance there is between us and it, between us and its author, between us and its readers, and between us and its time and context?

Well, plenty—and this book will show you how to do it. It will show you how to delve deeper into the historical texts that are often read in the history of mathematics—what things to look for and what questions to ask. And it will show you how to discover from historical texts—and about them—the things you want to know about the history of mathematics. How to read, in other words.

Chapter 1 will show you how to think about *what a historical piece of mathematics really says*, and how to find out about the processes of translation and interpretation it may have gone through before it reached us.

Chapter 2 will introduce you to the *author* of a text, and show you how to find and choose sources of information about the time and place in which it was written.

In *Chapter 3* you will learn about the *physical objects*—books, but not only books—in which historical mathematical texts are found, and what you can learn from them.

Chapter 4 will introduce you to the *people who read* historical mathematics when it was new, and will help you to ask and answer questions about who was meant to read a mathematical work, who really read it, and why it matters.

Finally *Chapter 5* will show you how to approach the elusive idea of *"significant" historical mathematics*, and help you to think about why we choose to read some historical writings and not others.

This book will, I hope, give students the skills and the confidence to tackle a course on the history of mathematics in which reading historical writings is required. It will also give general readers—or students taking a course that relies on a modern textbook rather than the historical writings themselves—the nerve to pick up a historical mathematics book or an anthology and have a go at making sense of it.

The history of mathematics is, as much as anything, a set of skills—ways of reading a particular kind of text. Historians of mathematics deal with a unique kind of writings, and they use a unique combination of skills to make sense of them. This book can begin to make you into a historian of mathematics.

The book uses specific examples to guide you through the experience of looking at a historical mathematical text, from thinking *what on earth was that?* to asking and answering detailed questions about what it is and what it means. The examples come from the mathematics of Europe between about 1550 and 1900—the historical mathematics that I happen to be most familiar with—though I hope the historical lessons will carry over to other cultures and periods. I've tried to provide some examples that are well-known classics and some that are a bit more unusual. The actual mathematics contains some undergraduate-level material (a little analysis and group theory, for instance), but much of the discussion doesn't require you to follow the mathematics in detail.

Preface

From time to time I'll invite you to stop reading and think through a particular point yourself before going on. These places are marked with the icon:

Pause for thought

Summaries set out in boxes provide handy reminders of the key techniques you have learned in each section and chapter, and in some cases give you quick help with particular situations.

Finally, at the end of each chapter there is a set of points for you to think about, taking the material further in various ways. These can be approached in many different ways, and there's no "right" answer to any of them: in a study setting they might be used for class dicussion or essay writing, but they could just as well be springboards to further thought and reading for the fun of it.

Acknowledgments

This book has grown out of my own experience in teaching the history of mathematics at the University of Oxford, and I am grateful to all of my students and colleagues there who have taught me—knowingly or otherwise—how to read historical mathematics. I would also like to thank the editor and director of *Contemporary Review* for his kind permission to reproduce a section from Oliver Lodge's article, "Einstein's Real Achievement," which appeared in the *Fortnightly Review* on 1 September 1921. This excerpt is reprinted in question 7 at the end of chapter 4 of this volume.

HOW TO READ
Historical Mathematics

CHAPTER I

What Does It Say?

When the cube and the things together
Are equal to some discrete number,
Find two other numbers differing in this one.
Then . . . their product should always be equal
Exactly to the cube of a third of the things.
The remainder then as a general rule
Of their cube roots subtracted
Will be equal to your principal thing.

—From Niccolò Tartaglia's account of the solutions
to the cubic equation (1539) in Fauvel and Gray,
The History of Mathematics: A Reader, pp. 255–56.

That's quite a mouthful. In your study of the history of mathematics, you'll quite often come across things like this. They can be baffling at first sight. On the other hand, the same piece of mathematics might be presented like this:

To solve $x^3 + cx = d$,
find u, v such that $u - v = d$ and $uv = (c/3)^3$.
Then $x = \sqrt[3]{u} - \sqrt[3]{v}$.

This looks much more straightforward: it's in a mathematical language which we can understand without much difficulty, and we can easily check whether it is true or not.

But it's not really obvious that the two versions say the same thing. Let's look in detail and see if we can trace how you get from one to the other. Before we start, pause for a moment and see how much of it you can make out yourself.

←Pause for thought

How far did you get? Give yourself a pat on the back if you managed to translate all eight lines into algebra and got something that made sense. Here's how it goes.

When the cube and things together

That's pretty cryptic, for a start. But I've told you that this is about solving cubic equations, so it's fair to assume that there's an unknown quantity—call it x—involved, and that "the cube" means x^3.

What about these "things"? Well, if this is a cubic equation, they can only be (1) a multiple of x^2, (2) a multiple of x, or (3) a constant. If Tartaglia meant a multiple of x^2, he would surely say something about "squares" or "the square," so we can rule out (1). There seems to be no way to tell whether he means a multiple of x or a constant for the moment, so let's leave that and look at the next line.

Are equal to some discrete number,

That makes things a bit clearer. "Some discrete number" sounds pretty much like a constant—let's call it d. That means that "things" is most likely a multiple of x, not another constant. Let's call it cx. So putting the first two lines together gives us this: "x^3 and cx together are equal to d." Or, to put it another way: $x^3 + cx = d$.

We're getting somewhere. The first two lines state the problem; the rest of the quote presumably tells us how to solve it.

Find two other numbers differing in this one.

Suddenly we're lost again. Find two numbers—find u and v, say—differing in "this one." This what? Tartaglia means "this number": that is, the "discrete number" from the previous line, the constant that we called d. So this line means "find u and v differing by d" or "find u, v such that $u - v = d$."

Then . . . their product should always be equal
Exactly to the cube of a third of the things.

"Their product" is the product of u and v. It's meant to be equal to "the cube of a third of the things." The last time the word "things" was mentioned it meant cx. Here that would give us $uv = (cx/3)^3$, right?

Wait a moment. If x is our unknown, we can't have it in our definition of u and v. What else can "things" mean?

Perhaps it means the coefficient: not cx but just c. That gives us $uv = (c/3)^3$, which makes a lot more sense. Now,

The remainder then as a general rule
Of their cube roots subtracted

"The remainder . . . of their cube roots subtracted"—this has to mean "the remainder when their cube roots are subtracted *from each other*." If we subtracted them from anything else, we wouldn't get one remainder, but two. So these lines mean $\sqrt[3]{u} - \sqrt[3]{v}$.

Will be equal to your principal thing.

There's no prize for guessing that "your principal thing" is the unknown quantity we are looking for, x. So these final lines mean $x = \sqrt[3]{u} - \sqrt[3]{v}$.

If you go back and look at what we've done, you'll see that the first two lines tell us that $x^3 + cx = d$; then the next six lines go on to tell us how to solve this equation: in line 3 we're told to find u and v such that $u - v = d$, and in lines 4 and 5 Tartaglia says we must also have $uv = (c/3)^3$. Then, in lines 6 through 8 he reveals that this gives us a solution: $x = \sqrt[3]{u} - \sqrt[3]{v}$.

In Modern Terms

If you're faced with a piece of historical mathematics that's not in modern notation, you can . . .

Work through it and translate it into modern terms.

If there are "quantities" or "numbers," make them x's and y's.

If there are squares and cubes, addition, multiplication, etc., write them out using algebra.

If there are things you can't work out, move on. Perhaps the next few lines will make it clear.

You could also . . .

Find a modern version of this particular passage and compare it with your own modern version.

Find a modern version of the same mathematical result and compare it with your own modern version.

You've just seen how to translate sixteenth-century words into modern algebra: not a trivial task, but not an impossible one either. If you want some practice, there's another example similar to this one at the end of the chapter.

Here's another example, with some harder mathematics in it.

> Quantities . . . which in any finite time constantly tend to equality, and which before the end of that time approach so close to one another that their difference is less than any given quantity, become ultimately equal.
>
> —Isaac Newton, *Philosophiæ naturalis principia mathematica*, Book 1, Lemma 1, translated by I. Bernard Cohen and Anne Whitman, 1999.

Once again, you might like to pause before reading on and have a try at translating this passage into modern notation, just as we did with the piece from Tartaglia. See how far you can get, and don't worry if you get stuck.

Pause for thought

Here's how we might translate this passage into modern notation.

"Quantities," the extract begins. Newton is talking about two quantities: let's call them X and Y. They change over time, and we're interested in their behavior over time, so we'll consider them functions $X(t)$ and $Y(t)$ of time t. In particular, we are interested in their behavior "in a given time." Assuming that time is finite, we can call it the period from $t = 0$ to $t = t_1$.

During that time, Newton says, the two quantities "constantly tend to equality." That means the difference between them always gets smaller; in other words, $|X(t) - Y(t)|$ is decreasing during our time interval.

Next, Newton gives a second condition on the two quantities, a more demanding one. During the specified time period they "approach so close to one another that their difference is less than any given quantity." If we call the "given quantity" ε, what that means is that sooner or later the difference will be less than ε. That is to say, we can find a time t' for which $|X(t') - Y(t')| < \varepsilon$. And that's true for *any* "given quantity": any value of ε. In modern terms, for all ε there exists t' in $[0, t_1]$ such that $|X(t') - Y(t')| < \varepsilon$.

The last part of Newton's sentence tells us the consequence of these two conditions. If the conditions are met, he says, the two quantities "become ultimately equal." "Equal" obviously means that $X = Y$; "ultimately" means that this happens at the end of the time interval we're considering, at $t = t_1$. So, $X(t_1) = Y(t_1)$.

So we've found that Newton's statement can be rewritten in modern notation like this:

Given $X(t)$ and $Y(t)$ with $|X(t) - Y(t)|$ decreasing from $t = 0$
to $t = t_1$,
if $\forall \varepsilon \, \exists t' \in [0, t_1]$ such that $|X(t') - Y(t')| < \varepsilon$,
then $X(t_1) = Y(t_1)$.

A bit harder than Tartaglia, but again, not an impossible task.

Spotting the difference

But now let's mess things up. Do you think this modern version is really—strictly—equivalent to what Newton wrote?

Be as picky as you can, and see if you can find some places where the two are not quite the same. Go back through our process of translation if you want, and check whether everything is absolutely watertight. You might find that it's not: you might have noticed even as we made our translation that we were introducing some changes that are not just matters of notation or of style.

Can you spot any differences? You might get some hints by comparing our modern version of Newton with the definition of a limit that you find in a modern textbook.

Pause for thought

What points did you come up with? These are really picky things, and I'd like to consider just four of them, though you might have found more than that.

First: What *exactly* does Newton mean by "constantly tend"? We've said that $|X(t) - Y(t)|$ is "decreasing," but for us that could mean that $X(t) - Y(t)$ is constant in places, or even constant everywhere, making $X(t)$ and $Y(t)$ equal across the whole interval. I doubt that's what Newton means—"constantly tend" gives the impression of things changing, not staying the same. Perhaps the idea of two quantities "constantly" tending to become equal might be better expressed by saying explicity that (1) they're not equal to begin with and (2) the difference between them is strictly (i.e., always) decreasing. Should we put that into our modern version?

Second: What exactly does Newton mean by a "quantity"? A real number? Yes and no. When were the real numbers first rigorously defined? Not until a long time after Newton: the late 1800s, in fact. That doesn't stop him from intuitively using them, and it seems obvious that he's thinking about quantities which change *continuously* here. On the other hand, it's not clear that the result is still true if the "quantities" and the "given quantity" are rationals, yet it seems a bit cheeky to foist onto Newton the condition that everything in sight is a real number when he wouldn't actually have known precisely what that meant. Should we have said that X, Y, t, ε, and t' are all real numbers, or shouldn't we?

Third: Can ε be negative? No, it can't; that wouldn't make sense—and it can't be zero either. So if we want our modern version to be strictly rigorous, we would have to specify that $\varepsilon > 0$. Newton doesn't say that; he just says "any given quantity" without pointing out that it has to be a positive nonzero quantity. Maybe for him a "quantity" is necessarily positive. Maybe he just thought it was obvious. Should we leave our version as it is, or should we put in that $\varepsilon > 0$?

Fourth: These are not the only differences you might have spotted, but it's time to move on—suppose that $X(t) = 1/(t_1 - t)$ and $Y(t) = X(t) - t_1 + t$. Think about it for a moment. The difference between them behaves as it should: it's $t_1 - t$ in the interval, and it satisfies the second condition too. But neither X nor Y is defined at t_1, so the result we're interested in—$X(t_1) = Y(t_1)$—is false in this case. What can we do?

There are ways to fix this up—we could add a condition that X and Y be bounded in the interval, for example, or that they be defined at t_1—but once again we'd be putting in something that Newton didn't say. Apparently he either didn't know about badly behaved functions like these or didn't care—or maybe he just thought it was obvious that they were not what he was talking about. Is it better to fix our modern version so it excludes cases like these, or to leave it closer to what's in Newton's version?

For each of the changes I've suggested making to our modern version, you could argue it either way, and I'll leave it up to you to decide what's best to do. The point is that there *are* some differences between what Newton said and our translation of it into modern mathematical language, and it's not easy to eradicate them without straying from what Newton really wrote. We'll come back to this again and again later in this chapter and throughout this book.

You probably found this exercise a bit picky. But think what we've learned about Newton's mathematics that we might have missed otherwise. His idea of "constantly tending" says more than our idea of "decreasing." When he says "quantities," he might be talking about the real numbers, but we can't really be sure exactly what he has in mind. He assumes that "any given quantity" is greater than zero, without saying so. And he assumes that his "quantities" are finite and generally well behaved in the interval he considers.

You've now seen two ways of finding out about a piece of old mathematics. The first is to translate it into modern terms; the second is to look closely at how the translation isn't *exactly* the same as the original. You can practice them as you go through the rest of this book, and on any other bits of historical mathematics you meet—and you'll find you can learn an awful lot about a piece of old mathematics like this. The box gives a summary of this second way. It gives a selection of questions you can ask to help you spot the difference between a modern

---

Spotting the Difference

Once you've made—or found—a modern version of a historical piece of mathematics, you can . . .

Look closely at the terms and the concepts, and how you've translated them: Is a "quantity" quite the same thing as a "number"? Is "tending" quite the same thing as "decreasing"?

Look closely at the assumptions you've made or would like to make: Do quantities need to be real, positive, bounded, well-defined? Does the author point those things out or not?

Look closely at the original argument and how precise it is: Is it actually correct? Are there counter-examples, extra conditions, special cases, etc., that the author doesn't bother to point out? There might even be mistakes, or gaps in the argument.

---

version and the original; but those are not the only questions you can ask. See if you can think of some others of your own. Throughout this book we'll see that being a historian of mathematics is about learning to ask your own questions.

Translation

There's something else, too. Newton didn't say "quantity," and Tartaglia didn't say "some discrete number" or anything like it.

They wrote in foreign languages: Newton in Latin and Tartaglia in Italian. Does that matter?

In a word, yes. Very often in the history of mathematics you'll read texts that have been translated from one language to another. A good translation will turn the words into English but should leave the mathematics and its notation more or less alone. If you suspect that the mathematics itself has been tampered with, or if you're lucky enough to be able to read the original language, you can always try to find a copy of the original version and have a look at it.

We'll spend some time in chapter 2 thinking about how to find sources for the history of mathematics in libraries and on the web. For the moment I'll just say that the first (Latin) edition of Newton's *Principia* is on at least one open-access website. In chapter 3 you'll learn about some of the other things you can find out by looking at the original source, and we'll also think in depth about some of the ways that sources can be tampered with before they get to us.

If you don't, or can't, have a look at the original, it might be possible to find more than one English translation of it. There aren't a great many pieces of historical mathematics that have been translated into English more than once, but not surprisingly Newton's *Principia* is one of them. Here's the same passage we looked at above, but from an older translation. (In fact, this was the first English translation of Newton to be published—in 1729—and it's on Google Books).[1]

> Quantities . . . which in any finite time converge continually to equality, and before the end of that time approach

---

[1] books.google.co.uk/books?id=Tm0FAAAAQAAJ.

nearer the one to the other than by any given difference,
become ultimately equal.

I think that's similar enough to be quite reassuring. The differences seem small: "converge continually" instead of "constantly tend"; "nearer the one to the other" instead of "so close to one another"; and "any given difference" instead of "any given quantity." Do any of these affect our translation into modern terms? I don't think they do. Do they affect what we can learn about what Newton was thinking? Again, I think not—though you might feel that this older version has a slightly stronger emphasis on motion and change: "converge" and "any given difference" rather than "tend" and "any given quantity." But if there is a difference, it's quite a subtle one.

That means we can probably rely on either of these translations to give us a pretty accurate idea of what Newton was saying, and unless you are willing and able to tackle the Latin that's as much as you can hope for. (If you're keen on these things, you might like to look at the introductions to Florian Cajori's English version of the *Principia* (1934) and the one by Bernard Cohen and Anne Whitman (1999): Why did the editors feel there was a need for a new version? Cohen and Whitman have quite a lot to say about the problems of translating the *Principia*. And if you *can* tackle the Latin, do. The original text is on the web on Gallica, and you can make up your own mind about exactly how it should be translated.[2])

---

[2] gallica.bnf.fr/ark:/12148/bpt6k3363w. Not to get bogged down in complications, you should know that this is the first edition, whereas the two English translations were based on the third edition, which is slightly different. How it's different is interesting too, but that's another story.

How they thought

When you become accustomed to reading pieces of historical mathematics, you'll find that you don't always need to make modern versions of them in order to see what they are about. And you might also start to feel that modern versions have their limits. You'd be right. While it can be very handy to have the basic mathematical contents of Newton's text—or whoever's—set out in modern algebra, the point of reading historical mathematics is often not just *what they said* but *how they said it*. After all, if you just wanted to see a definition of a limit or a solution to the cubic equation, you could look in a modern textbook.

All the differences we've looked at—the ways our modern version didn't say exactly what Newton said, and the fact that two different English translations of the passage aren't *exactly* the same—are ways you can learn more about a piece of historical mathematics. By looking at not just *what authors said* but also *how they said it*, you can start to learn about *how they thought*.

That sounds hard, but you've started to do it already, when you noticed that Newton says "quantity" when we say "real number," and that he didn't bother to point out some of the assumptions that we want to make explicit. And so far we haven't even mentioned one of the the biggest differences between our version of Newton and the original. That difference is the notation—or rather the fact that our version *has* some mathematical notation and Newton's doesn't; it just has words. You've seen that it can be difficult just to understand a mathematical statement written down in words alone. Imagine how hard it must be to make mathematical discoveries like that, without using any algebraic notation. The solutions to the cubic and the quartic equations were first worked out like that, for example—it's very hard to imagine how they did it. There's an example at the end of this chapter where you can learn more about doing mathematics without notation.

What does the lack of notation mean for the passage from Newton? We already noticed that Newton says "quantities" when we say X, and we would like to make X a real number. More than that, in what Newton says, there are no variables and no functions at all, even though our modern version has both. When we think about this piece of mathematics, we probably imagine a graph of the functions X and Y against the variable t, with two curves that approach one another and eventually meet. But Newton was probably imagining exactly what he says—a quantity which changes over time. Maybe he was thinking of moving bodies whose positions approached one another over time; maybe their *speeds* approached one another; maybe he was thinking of a geometric construction where two *lengths* gradually became the same as he changed a third length (for example, two sides of a triangle: however they start out, if you make the third side decrease to zero, they will become equal).

The word "function" is a notoriously complicated term in historical mathematics. When Leonhard Euler wrote "function" in 1748 he meant, by his own definition, "an analytic expression composed in any way whatsoever of the variable quantity and numbers or constant quantities." When he wrote "function" in 1755, he meant, again by his own definition at the time, that "when quantities depend on others in such a way that [the former] undergo changes themselves when [the latter] change, then [the former] are called functions of [the latter]. . . ." When we say it, there are a few different definitions, depending on context, but none of them is really the same as either of Euler's.[3] This is a good example of how words and concepts don't stay still over time—and how looking at them can show us how historical mathematicians thought.

Something that can often help in this situation is to look look at how the author uses his words and concepts later on in his book or paper. When you see the mathematical words and concepts in use, it can become a lot clearer what they are about; looking ahead often allows you to figure out what the author meant originally. In the case of Newton, if we look ahead in the *Principia,* we find him, again and again, using his idea of a limit in geometric situations where he is interested in the relationships between quantities—lengths or areas—that change over time. For him this piece of mathematics is about changing geometric quantities and what we can say about them when they gradually become equal.

This difference—that we're thinking about functions or a graph and he's thinking about a changing quantity—is a difference we can't fix by modifying our modern version. His concepts are just different from ours. This can be confusing, but it's also helpful to us as historians because it gives us a window into how Newton was thinking—not just what he said and how he said it, but how he thought.

You'll have plenty of chances in this book to practice thinking about this. Old mathematics might use different concepts from the ones we'd use in the same situation—like Newton

---

[3] There are more examples in Victor Katz's *History of Mathematics* on p. 724 of the second edition.

talking about "quantities" where we'd talk about "real numbers." It might use concepts that you've never come across before—then you would have to see if you could figure out what the author meant, or if the same author gave a definition that would help you. Or it might use concepts that seem familiar, but use them in surprising ways or with surprising meanings—like Euler's different meanings of "function."

By noticing any of these things—even if they seem a bit confusing—you learn something about the historical mathematics and its author. You learn not just *what mathematics it has in it* or *how that mathematics is expressed*, but you also learn something about *how a historical mathematician thought*.

How They Thought

To learn about how a historical mathematician thought, you can ask . . .

What notation does the text use? What words? What concepts? How are these different from what you would use in the same situation?

Does it use words or concepts you don't recognize? Can you work out what they mean, or find out what they mean from the author's definitions?

Does it use familiar words or notation, but with different meanings from what you would expect? Again, can you work out or find out exactly what the author means by them?

Conclusion

You've now seen three ways to learn about a piece of historical mathematics: three kinds of questions you can ask yourself that will help you make sense of *what the author is saying, how the author says it*, and *how the author is thinking.* See the box for a summary of what you've learned in this chapter.

There will be times, too, when you *don't* have the original piece of mathematics in front of you: just a modern version of it. This book is about reading historical mathematics, but what you've learned so far might also help you to read the modern version in that situation—and think about what it does or doesn't tell you about the original author and the original ideas.

What Does It Say?

When you're reading a piece of historical mathematics, you can ask . . .

What does it say? What mathematics is in it? Can you translate it into modern terms?

How does it say it? What are the differences between the modern version and the original, and what do they tell you?

How is the author thinking? How are the terms and concepts, and the approach to this piece of mathematics, different from what yours would be?

I'll leave that for you to think about because now we're going to move on to consider mathematical authors themselves.

To think about

(1) Translate the following into modern algebra:

> When the squares and the things are equal to a number, first you must reduce all the equation to one square, that is if there is less than one square you must equally restore and make good. And if there is more than one square you must reduce to one square, and reducing is done by dividing the whole of the equation by the amount of the squares. And when you have done this, halve the things, and multiply one half by itself. The number is added to this product, and the root of this sum minus the half of the things is the value of the thing required.
>
> —Luca Pacioli, *Summa de arithmetica . . .* (1494),
>   in Fauvel and Gray, p. 251.

(2) Find a statement of Newton's Second Law of Motion in a modern textbook. Compare it with this, Newton's original statement of the law (Cohen and Whitman's 1999 translation):

> A change in motion is proportional to the motive force impressed and takes place along the straight line in which that force is impressed.

You might like to think about the concepts Newton does and doesn't use, and you might also want to see if you can find out exactly what he means by the words "motion" and "motive force."

(3) Consider this version of Newton's Lemma 1, the piece of the *Principia* that we discussed above:

> If two quantities $X(t)$ and $Y(t)$, depending continuously on "time" t, and neither of which vanishes in the range $t_0 < t < \infty$, are such that
>
> $$\lim_{t \to t_1} \left[\frac{X(t)}{Y(t)} \right] \to 1,$$
>
> for some assigned $t = t_1$, then
>
> $$X(t_1) = Y(t_1).$$
>
> —S. Chandrasekhar, *Newton's Principia for the Common Reader*, 1995.

In what ways does this say more than our modernized version or than Newton's own words? In what ways does it say less? Do you think it is preferable to ours? Why?

(4) Discuss the following:

> If a glass tube, 36 inches long, closed at top, be sunk perpendicularly into water, till its lower or open end be 30 inches below the surface of the water; how high will the water rise within the tube[,] the quicksilver in the common barometer at the same time standing at 29½ inches?
>
> —Mathematical question no. 1041 by Mr. John Ryley, of Leeds, *The Ladies' Diary*, 1798.

Let l = 36 inches the length of the tube, b = 30 inches the part immersed, x = the height of water in the tube, and f = 413 inches, the height of a column of water equal to the pressure of the atmosphere, when the quicksilver stands at 29½ inches. Then, since the spaces occupied by the same quantity of air, are reciprocally as the compressing forces, it will be, as $l - x : l :: f : \dfrac{lf}{l-x}$ = force of the air in $l - x$; hence $\dfrac{lf}{l-x} + x = b + f$, and x = 2.2654115 inches.

—Answer by Miss Maria Middleton, Eden,
near Durham, *The Ladies' Diary*, 1799.

(You might like to think about the notation—can you work out what :: means?—as well as the physical assumptions the two writers make, any gaps in the mathematical proof, and the level of precision of the final answer.)

CHAPTER 2

///

How Was It Written?

My dear friend, I have done several new things in analysis.

Some concern the theory of equations; others, integral functions.

In the theory of equations I have found out in which cases the equations are solvable by radicals, which has given me the occasion to deepen the theory and to describe all the transformations admitted by an equation, even when it is not solvable by radicals.

One could make three memoirs of all that.

The first is written, and despite what Poisson has said about it, I hold it aloft with the corrections that I have made.

The second contains quite curious applications of the theory of equations. Here is a summary of the most important things.

(1) Following propositions II and III of the first memoir one sees a great difference between adjoining to an equation one or all the roots of an auxiliary equation.

In both cases the group of the equation is partitioned by the adjunction into groups such that one passes from one to the other by means of the same substitution; but it is only in the second case that it is certain that these groups have the same substitutions. This is called a "proper decomposition."

In other words, when a group G contains another group H, then the group G can be divided into groups that are obtained by performing the same substitution on the permutations of H, so that

$$G = H + HS + HS' + \ldots$$

It can also be divided into groups with the same substitutions so that

$$G = H + TH + T'H + \ldots$$

These two kinds of decomposition do not ordinarily coincide. When they do, the decomposition is said to be "proper."

It is easy to see that when the group of an equation does not admit any proper decomposition then one can transform it all one wants; the groups of the transformed equations will always have the same number of permutations.

If on the contrary, the new group admits a proper decomposition, so that it is divided into M groups of N permutations, then one can solve the given equation by means of two equations, one having a group of M permutations, and the other one of N permutations.

Thus, when one has exhausted in the group of an equation all the possibilities of proper decomposition, then one has groups that one can transform but that always have the same number of permutations.

If each of these groups has a prime number of permutations, then the equation will be solvable by radicals; otherwise, not.

The smallest number of permutations an indecomposable group can have, when this number is not prime, is $5 \cdot 4 \cdot 3$.

(2) The simplest decompositions are those which arise by Gauss's method.

As these decompositions are obvious even in the actual form of the group of the equation, it is pointless to spend a long time on them [...].

You will print this letter in the *Revue Encyclopédique*.

I have often in my life dared to advance propositions I was not sure of; but everything I have written here has been in my head for over a year, and it is too much in my interest not to deceive myself so that someone could suspect me of stating theorems of which I didn't have a complete proof.

You will publicly beg Jacobi or Gauss to give their opinion not of the truth but of the importance of the theorems.

After this, there will, I hope, be people who will find it to their advantage to decipher all this mess.

What on earth is this? Who wrote it? When? Where? Why? *How can we learn more?* In this chapter you'll find out.

This passage tells you certain things—you might already know them—about the decomposition of groups. The mathematical language and the concepts seem to be fairly similar to ours—if you did a translation into modern terms, the way you learned to in chapter 1, there wouldn't really be much to do. But I've deliberately not told you who wrote it or when or why, and without knowing those things you probably find it quite hard to know what to think of it.

So, what can you work out? Looking just at the passage above, what does it tell you about the author, and when and where it was written? You may be thinking, "not much." But look really closely at the passage: "squeeze" it for information. The author doesn't deliberately set out to tell us about himself (he's a man, by the way) or his world, but he actually tells us quite a lot about both.

For example, is this a member of the mathematical establishment, or an outsider? Is he young, or old? Does he get on well with other mathematicians, or not? Is he writing for publication, or something more informal? Does he seem calm, or in a hurry? What about the date when this was written—what can you figure out from the references to other mathematicians whom the author has apparently met or corresponded with? And what about the country where he's writing? There is a clue to that, too. Be a detective: go through the passage carefully and see just how many clues you can find—and what you can deduce from them.

←Pause for thought

Detective work

How much did you find out? There's an awful lot there. Here are some of the things that I found—but you might have found others.

The writer (let's call him X) mentions Poisson, Gauss, and Jacobi as, apparently, active mathematicians—and he implies that Gauss and Jacobi are alive at the time of writing and Poisson had been alive at least fairly recently. If you know their dates—or quickly check them—that tells you a lot about the date of this piece. (Siméon-Denis Poisson lived from 1781 to 1840, Carl Friedrich Gauss from 1777 to 1855, and Carl Jacobi from 1804 to 1851.) It can't be later than 1851, when Jacobi died, and it can't be much later than 1840, when Poisson died. On the other hand it can't really be earlier than about 1825 because of how young that would make Jacobi. That gives us about 1825–1840 as the date of this passage.

What about where this was written? The one journal X mentions is a French one, so your best guess is presumably that he's French himself. If that's correct, then he was probably writing in France; but you can't be completely sure of that—there's nothing to stop this from having been written somewhere else (Niels Abel was Norwegian, for example, but he wrote and published in both French and German). If X *was* French, and writing in French, then this is a translation. After what we said in the last chapter about texts sometimes being tampered with when they're translated, you might wonder whether the group notation used is what X really wrote. For this exercise, though, that doesn't matter.

The comment about Poisson suggests X had had some sort of disagreement with him: Poisson has apparently "said" (maybe that means "written") something negative about a paper by X. That in turn implies that X is at least trying to get his mathematics published in the form of papers in journals. He's someone who aspires to be taken seriously as a mathematician, but his defensive tone suggests he's not doing too well at that. It's possible that this is an older man who is coming to mathematical research late in his life—or an amateur trying to break into professional circles—or a young man struggling to be taken seriously. There's nothing obvious here to distinguish among those three possibilities, although you might just suspect that he sounds like a fiery young man rather than a cantankerous older one.

X refers to "this letter"—so this text is from a letter, apparently written to an individual; that's not where you'd normally expect to find reports of new mathematical discoveries. And it presents a lot of mathematical results in an extremely compressed form; X seems to have been in a hurry. He also mentions that "it is pointless to spend a long time" on certain decompositions of groups, which gives the same impression. And since he has had all of this mathematics "in my head" for the last year, why is he writing it up in a brief letter instead of actually writing and publishing the second and third "memoirs" which he mentions? All this seems to imply he's under a lot of pressure. Circumstances over the previous year have prevented him from writing—does he have a demanding job? has he been traveling? ill? in prison? there's no way to tell—and even now he only has time to jot down a brief summary. When he says "you will print this letter," and mentions that it will be for other people to "decipher all this mess," he implies that this is the only version of these results he anticipates writing, perhaps ever. So, whatever his problems are, he thinks they're going to continue. Maybe he's about to go on a long journey—maybe he's very ill—maybe he just doesn't forsee any end to the current pressures on his time from work, family, or other commitments. (France in 1825–1840, as you might know, had a major regime change in 1830, bringing in what is called the July Monarchy. You might possibly guess that that had something to do with the fact that X is in such a hurry and seems uncertain about his own future.)

The final sentences tell us something more about X's opinion of his own mathematical work. They're complicated to interpret. He says he has often "dared to advance opinions I was not sure of." Does that mean he's admitting he has made mistakes in the past? Probably. But this time he's confident of his results. He wants Jacobi or Gauss to be asked about the results' *importance*—not their *truth*—and he insists that his letter must be printed. His comment about people who will "decipher all this mess" also implies that it'll be worth their while to do so. Most people don't think their private letters deserve the attention of major mathematicians, or that they deserve putting in order and publishing. X does—he thinks his results are valuable.

Let's put all our guesses together. X is French, and he's writing in France around 1830. He's young and confident—he's been criticized for sloppiness in the past, but this time he's sure that what he's got is right and important. He's either dying or entering circumstances in which he'll never again be able to spend time on mathematics, despite his belief in its importance and in his own abilities, and he's hastily writing things up in a letter where he asks the recipient to get opinions from Gauss and Jacobi and to publish the results.

Maybe I've overdone it a bit—after all, I already know who wrote this—but I hope you've seen how a little lateral thinking and detective work can help you here. You can learn a lot about the author and the circumstances just from a single piece of mathematical writing—you can, so to speak, get a lot of context from a text.

Of course, you might already have known some of the answers. The passage is from the final letter of Evariste Galois (1811–1832),[1] and it features in one of the best-known stories in the history of mathematics. Galois fought a duel on 30 May 1832 which led to his death the following day. During the night before he went out to fight, he wrote this letter to his friend

---

[1] Printed in *Journal de Liouville* 11 (1846), 408–15; reprinted in Evariste Galois, *Œuvres mathématiques* (Sceaux, 1989), 408–15; the translation is from D. E. Smith's *Source Book in Mathematics* (Dover, 1959; first published by McGraw-Hill, 1929), pp. 278–85.

Detective Work

If you only have the text itself to work from, you can . . .

Try to work out the date: it might refer to mathematicians, discoveries, or other things that might help.

Try to work out the place of writing: it might refer to people, institutions, publications, that would give you clues.

Think about the author: the text might say that the author is old or young, or mention a place of work.

Think about the author's circumstances: the text might refer to other publications, conversations, or responses from other mathematicians.

Do some lateral thinking: for example, if it's a facsimile, think about the style of printing. You might be able to guess which century it's from, by comparing it with other pictures of old books. Clumsy notation or obvious mistakes might also suggest the author was doing something new which the printers had trouble with.

"Squeeze" the text: be a detective. Quite small clues can lead to quite large insights both about the author and about what the author was thinking or doing when the piece was written.

Think sideways, and really see how much you can work out that the author doesn't mean to tell you.

Auguste Chevalier, outlining his main mathematical findings and adding some corrections and clarifications to one of his (unpublished) papers. Galois had earlier sent a paper—now known as the *premier mémoire*—to the French Academy, and it had been returned by Poisson with negative comments. He had spent time in prison during 1831 on charges relating to his political opinions—which obviously had disrupted his mathematical work.

So our deductions were pretty good, on the whole. We didn't quite manage to tell which street he lived on and what he had for breakfast, but we did work out plenty of things about this author and his life and times without having to spend time with other books.

The Galois letter is quite an extreme example—most mathematics isn't written the night before a fatal duel. But by thinking about it, you've just begun to learn one of the main skills of historians: how to work out context from a text. This is something you can always do with a piece of historical mathematics—there are some ideas at the end of this chapter for you to practice with, and you might like to try out this skill on some of the other examples in this book. The box summarizes how to do it—but remember that there are no precise rules for this kind of detective work.

In the library and on the web

This chapter is about the authors of mathematical writings and how you can find out about them. So far, you've learned how to work out from a mathematical text itself things which the author probably didn't mean to tell you. What other sources of information can you use to find out about an author?

The short answer is the library and the web. Let's start by going to the library and see what we can find there about our author. For this we need a really big-name example to start with, so let's go back to Newton. What can we find?

If you look on the shelves of a college or public library, you'll probably find quite a few books about Newton. They might be scattered about: some with the mathematics books, some in the history section or the history of science section—or the history of mathematics section, if you're lucky enough to have one. There are some books that you'll find by doing a keyword search in the electronic catalog. There are others you might only notice by actually looking at the shelves.

Start with works Newton *wrote himself*—there's the *Principia mathematica*, which has three English versions (the translation by Andrew Motte from the eighteenth century, Florian Cajori's revision of it in 1934, and the new translation by Bernard Cohen and Anne Whitman published in 1999). There's his *Opticks*, which comes in quite an old edition. His other works, for instance his *Universal Arithmetick*, are a bit harder to find in most libraries.

There are also things Newton *wrote but didn't publish*: his letters, for instance, which fill seven volumes in their modern edition, and his *Mathematical Papers*, which fill eight volumes. A large history of science library might well have both.

All these things, writings which were created "at the time," so to speak, and are *direct* sources of information about a historical period even if they were published later, are called *primary sources* (there's a box below which summarizes the terms we'll use).

Second, there are books *about Newton*. Biographies, for a start—you'll probably find more than one: there's a big one by Richard Westfall, *Never at Rest*, and smaller ones by David Berlinski, Frank Manuel, and others. There are also books *about his writings*: for example, Subrahmanyan Chandrasekhar and Dana Densmore have both written book-length versions of parts of the *Principia* using modern notation, and Bernard Cohen wrote an *Introduction to the Principia*. And there are more general works which comment on Newton's significance, like Cohen's *The Newtonian Revolution*.

You'll notice, flicking through them, that all of these books have bibliographies in the back listing many, many more works about Newton, from popular guides to specialized research papers.

All of these materials, things written "after the time," and giving us *indirect* information about a historical period, are called *secondary sources*.

There is more, though. Any general book on the history of mathematics probably has a section about Newton: he gets nearly 20 pages in Victor Katz's *History of Mathematics*, for example. Many encyclopedias have articles about Newton, and some have more than one—*Encyclopedia Britannica* has entries on "Newton's rings," "Newton's law of gravitation," "Newton's laws of motion," "Newton and infinite series," and so on. There are shorter, more popular biographies like the ones by Michael White and by James Gleick.

I'm going to call these things "tertiary" sources—though that is not a standard term. Like secondary sources, they're written after the time they refer to. They are based on secondary sources, just as secondary sources are based on primary sources. While they can be excellent, there can also sometimes be problems with the quality of the information they contain—that's understandable since they are twice-removed from the primary sources, which are our only direct sources of information about the historical period in question.

You might be wondering what comes after tertiary sources. Some encyclopedia or magazine articles are based on other encyclopedia articles. Should we call them "quaternary" sources? I think we should stop at "tertiary" and use that term for anything not based on the primary sources. In any case, there are no rigid lines between the three categories. Many writers mix information from primary and secondary sources, and it is up to you to decide whether an author is mainly studying the original sources or mainly repackaging other historians' ideas.

Primary, secondary, . . . tertiary?

Primary sources

—are our only *direct* sources of information about a historical period;

—are created "at the time," like the pieces of historical mathematics we'll meet in this book, or like someone's diary or letters;

—need not be written texts: they can be artefacts, like items of clothing; or buildings, photographs, or sound recordings; or even someone's memories.

Secondary sources

—are one step away from the historical events they tell us about;

—are produced after the events they relate to;

—are based on primary sources and normally come with a list of the primary sources the author has consulted;

—include many academic books and journal articles, and some encyclopedia articles (like those in the *Oxford Dictionary of National Biography*, for example).

"Tertiary" sources

—are what comes after secondary sources;

—are produced after the events they relate to;

—are based on secondary sources, or even on other tertiary sources, and are written by authors who probably have not studied the primary sources;

—are two steps away from the historical events they tell us about;

—don't always tell us where their information comes from;

—include most encyclopedia articles, website articles, and books aimed at a wide audience.

So, for a big name like Newton there is a whole array of different sources of information just in the library—we haven't even looked on the web yet. You don't have time to read everything: How do you decide which sources to look at? Which ones will tell you what you want to know? Which can you trust?

The division into primary, secondary, and tertiary sources can help you choose. It tells you something about the amount of "transmission error" you can expect. Reading the primary sources is the best you can do in that respect—but sometimes you need more than that. If you're in a hurry, you might go to an encyclopedia article, which offers basic information quickly. But with a tertiary source like an encyclopedia article there can be a lot of other people's opinions and interpretations mixed in with the historical information. We all know about "spin," and when information is second or third hand, it can be factually accurate but still contain quite a lot of spin—and it can even be incomplete or misleading or contain errors copied from other secondary or tertiary sources. Unless you already know the subject in question, you might have no way to spot error or bias unless it's really very obvious. What these tertiary sources are good at, though, is giving you some preliminary orientation. They can help you find your way around a particular subject, or give you pointers to the secondary sources where you might find more detailed help.

If primary sources are too dense and confusing and tertiary sources are not detailed enough or, perhaps, reliable enough, the best route is often to compromise by going to secondary sources—biographies and other books about a historical figure or a historical period. These offer the views of someone who has actually seen the primary sources in question and can write authoritatively about them and the subject, but without needing to read the primary sources in bulk yourself.

What about the web? you're probably thinking. All three kinds of sources are plentiful on the web, and many of them are open-access. For Newton there's The Newton Project,[2] which puts up a lot of primary sources together with commentaries,

[2] www.newtonproject.sussex.ac.uk.

bibliographies, and essays—solid secondary sources. Some of his writings are also on Google Books or Gallica,[3] another open-access site with facsimiles of old books and journals—primary sources. And there are any number of encyclopedia articles on the web about Newton, and about all sorts of aspects of his life and writings—tertiary sources. The obvious places—the St. Andrews University history of mathematics site[4] and Wikipedia—can be good places to start, though of course they're not immune to the problems of all tertiary sources—but don't stop there, particularly with someone like Newton for whom there's so much available online.

Also, an institution might give you on-campus access to resources like Early English Books Online or Eighteenth-Century Collections Online.[5] Between them these two have digitized versions of all the books Newton published during his lifetime, as well as other Newton-related items like, say, Colin Maclaurin's 1742 *Treatise of Fluxions* (a primary source or a secondary source? What do you think?). You might have access to JSTOR[6] and other archives of journal articles, where a search will find research articles on many aspects of Newton's life and work. JSTOR, as it happens, also has digitizations of the complete back issues of *Philosophical Transactions*, the journal where Newton published some of his own work in the 1600s—so it's a source of both primary and secondary sources.

There are several different archives of electronic journals,

---

[3] gallica2.bnf.fr.
[4] www.gap-system.org/~history/index.html.
[5] eebo.chadwyck.com.
[6] www.jstor.org.

In The Library and On The Web

If you're looking for more information about a piece of historical mathematics or its author, you can . . .

Search in library catalogs.

Look in encyclopedias like *Britannica* or (if appropriate) the *Dictionary of National Biography* or *Dictionary of Scientific Biography*—use the bibliographies in these works as signposts to primary and secondary sources.

Go to the library and check the shelves (try the history section and the history of science section as well as the mathematics section).

Search the web.

Search specific online resources like Google Books, Early English Books Online, and JSTOR.

Select a few primary and secondary sources to help find out what you want to know: maybe one biography and one collection of the person's writings, for instance; or a couple of journal articles about aspects of the person's work.

and many universities now have centralized systems for managing access to them. If you're looking for recent journal articles, a search on Web of Science or Scopus[7] might also help. But the web moves fast, and by the time this book is printed there will probably be new tools, both open- and restricted-access, for

---

[7] apps.isiknowledge.com, www.scopus.com.

finding all these kinds of sources, and it will be up to you to discover them and use them.

Questions to ask

Now you've found a few sources, what do you do with them? What questions do you ask them? You've got the books on your desk or the documents on your desktop—it might be fun to just browse through them, but you'll probably learn more, and write a better essay later, if you have some idea of what you're looking for or what questions you're asking.

Try making a list of questions you could ask about Evariste Galois, the author of the extract above. Think about what you'd need to know to understand more about his mathematics, about why he did it and how he did it. Then think about where you might look—which kinds of sources?—for answers to your questions.

Pause for thought

You probably came up with quite a long list of questions quite quickly. There's almost no end to the things you can ask about a historical person. I'll walk you through one set of questions, and I hope you find it helpful. But I also hope you'll come up with your own ideas about what questions to ask and how to organize them.

Where and when? You can ask about Galois' biography—where and when did he live, where and when did he do most of his mathematical work, and where and when did he write the text you looked at above?

You can answer the first two questions from the *Dictionary of Scientific Biography*, or probably any other encyclopedia article about him: he lived all of his life in and around Paris, he was born in 1811, and he died in 1832. For the third question you could perhaps look in the introduction to this passage in Fauvel and Gray's *The History of Mathematics: A Reader*. It would also be answered in some encyclopedia articles. In fact, the letter was written on the night of 29 May 1832, in Paris—the day before Galois was shot and two days before he died.

Another way to find out about the writing of this text is to look in a biography of Galois. Searching Amazon for "Galois biography" is probably the quickest way to track one down, and you'll find that there's just one biography of him in English, by Laura Toti Rigatelli (it's on Google Books). There are also several novels about Galois, including *Whom the Gods Love*, by Leopold Infeld, and *The French Mathematician*, by Tom Petsinis. In French there's another biography, by Alexandre Astruc, and in German there are a biography and a novel, by Bernard Bychan and Bernd Klein, respectively. And that's not all.

This gets quite confusing quite fast, since there are various different stories about how Galois died—a duel or an incident stage-managed by one political group or another—and it's hard to figure out quickly which one is best supported by the evidence. In this situation it's fair to say "historians disagree . . . ," but with a closer look at one of the more scholarly sources you may be able to make your own judgment about which version of events is most convincing. What does come across if you look at more than one version is that you probably shouldn't take the brief biographies in encyclopedias at face value in this case. There's more biographical material, and a lot more about Galois, at www.galois-group.net and elsewhere on the web. Let's move on.

What and how? Now you know when and where Galois wrote this. But that doesn't seem to be very much. "Paris in the 1820s" is just a date and a place. What does it mean to say that Galois lived then? To put him in context, it would be good to know something about what was going on in that time and place. So, *what* was that time and place like for Galois? *How* did he experience it?

For example, what sort of family did he have? What was his education like? What jobs did he have? How was his life affected, helped, or disrupted by what was going on in France and Paris at the time? How might all of this have affected his work?

This is where it can really help to browse through a full-length biography if you can find one, or to look closely at a good encyclopedia article and chase down its references to political events and important people of the time. You can set out the main facts of Galois' life as answers to the questions we just asked. He was from a middle-class family; his father was a mayor, but he hanged himself when Galois was 17; Galois went to a good local school, but twice failed the entry examination for the École Polytechnique in Paris—which was effectively the leading Paris university at the time—and went on to study instead at the less prestigious École Normale. He didn't live long enough ever to have a job, and his nonmathematical activities seem to have consisted mainly of revolutionary politics that landed him in jail for part of 1831. If you've read about him in encyclopedias or a biography, you have probably also got some impression of him as a person—the shape of his life and what was important in it.

One of my reasons for using Galois as an example in this chapter is that all of these things *are* relevant to his mathematics. You can make an obvious connection between his revolutionary politics and his attitude in mathematics: wanting to overturn other people's ideas without always making his own as careful or as clear as others thought they should be. He was a hot-headed young man who got himself jailed and then fatally shot, and he was also a hot-headed mathematician who sent a paper to the Paris Academy of Sciences which Poisson claimed he couldn't even understand. The state of some of his mathematics— unfinished drafts, unpolished and even chaotic versions of his ideas—makes no sense unless you know he lived in turbulent times and took an active part in them, distracted from his mathematics even though he felt it was deeply important.

Even when it's not so obvious what the connections are between someone's life and mathematics, you can always make *some* connections. Busy or not busy, with a comfortable job or wandering about in exile—mathematics always gets done in real circumstances, and you'll know more about why the mathematics is written the way it is if you know roughly what those circumstances were.

Who and why? What you've just thought about is the *biographical* context for someone's work. There's also an *intellectual* context. *Who* were this person's colleagues? What else was going on in mathematics at the time? To put it another way, *why* was

Galois doing group theory and not, say, non-Euclidean geometry or fluid dynamics?

Once you ask those questions, it's obvious what some of the answers are. You can't do group theory before you've done (some) algebra, for example—so there's a logical order for some parts of the history of mathematics. On the other hand, once you've done some algebra you don't *have* to do group theory, let alone Galois theory. So the "why" question comes back again—why did Galois find this interesting, and not something else? (One answer is that if he hadn't we wouldn't be studying him—but we'll come back to that in chapter 5.)

So, you might want to know something about the general kind of mathematics that was being done at the time, and how Galois' work fitted in—or didn't. If he was a more sociable mathematician you might ask: Was he in a research institute which promoted pure mathematics? Was he in a group of colleagues who were all interested in group theory? Was he working with a collaborator who pushed him toward specific research questions? All these things can help you understand why a particular piece of mathematics was written, why it asks the questions it does, and why it asks them in the way that it does.

If he had lived longer, you might also ask how this piece of mathematics fitted into his career, and whether there were external reasons making him write—maybe his employer was hassling him to publish something, or maybe his students needed a better textbook.

In fact, Galois didn't really have any colleagues, as you'll have gathered from the sources you've looked at. He was 20 when he died, and he'd been expelled from college a year and a half before. He'd tried submitting his work to the Paris Academy, but it was rejected with comments he evidently resented.

He was perhaps more isolated than most of the mathematicians we study—but that didn't mean his work was out of touch with other people's. The extract you looked at above mentions the question of which equations are solvable by radicals, and the transformation of equations. Those are both topics on which you'll probably find out something from the index of a general book about the history of mathematics. We had a glimpse of the earlier history of equation solving in chapter 1. The fact that no solution to quintic equations was known had become a research problem in its own right by the second half of the eighteenth century, and Joseph-Louis Lagrange, for instance, had written about the theory of equations in 1771, trying to understand why the methods for solving cubic and quartic equations worked, and how they could—or couldn't—be adapted for higher-degree equations. An attempted proof of the unsolvability of the quintic had appeared in Italy in 1799, and Niels Henrik Abel had published his correct proof in 1824, with an expanded version appearing in August Crelle's German mathematics journal in 1826. His proof applied the theory of permutations to the set of roots of the equation.

Group theory has a shorter history than equation solving, but again you can find out more about it from the standard textbooks on the history of mathematics or by looking around for more detailed sources. As the story is often told, the concept of a group first emerged in Lagrange's work on permutations, in the context of the number of different values which an algebraic function would take if its variables were permuted. The concept was taken further by Augustin Louis Cauchy early in the nineteenth century, and Galois then took things still further, examining the permutation structures—groups—that would describe transformations of equations, and describing those groups in their own right.

So Galois was contributing to a subject that other mathematicians were interested in at the time. Maybe he was deliberately attacking a "cutting-edge" problem—and his comment about making ambitious statements which he can't fully justify tells you something about how he approached it.

The box summarizes some of these different kinds of questions that you can ask once you've found some useful sources of

---

Questions to Ask

Once you've found some useful sources you can ask . . .

Where did the author live? *When*? And *where* and *when* was the text you're looking at written?

What were that time and place like for this particular person? *What* was going on in politics or society? *What* would have been the "current affairs" of the time?

How did this person experience those things? Think about family, education, job(s), social position, and the kind of people this mathematician would have known. *How* was life affected, helped, or disrupted by the major events of the time?

Who were this person's colleagues? Who were the other people—friends, colleagues, students, editors—who were involved in making this particular piece of mathematical writing happen and getting it published (if it was)? And what about any enemies?

Why was this person working on this particular piece of mathematics? What else was going on in mathematics at the time? Think about the overall shape of the writer's career, and the colleagues and research environment involved in it. Was this piece of writing meant to help with teaching, to further a career, to satisfy the writer's curiosity or someone else's?

information. But remember that you can always come up with questions of your own too.

In chapter 1 you mostly thought about your own reading of a text and how to find mathematical meaning in it. In this chapter you've looked at the author: the author's motivations and context, and what you can learn about a mathematical writer both from the text at hand and from other sources (it's summarized in the box below). These two approaches—focusing on the reader or on the writer—are fundamental for doing history. One way to put it is to think: What do we get out of the text and why? What did the author put into it and why?

In the next chapter, we will think about books—the actual physical objects that contain historical mathematics.

How Was It Written?

To find out more about how, when, and by whom a piece of historical mathematics was written, you can . . .

Be a detective about the text itself: see how much you can learn from it about the author—the author's life, mathematics, and world.

Use your library: see what's on the shelves as well as searching in the catalog; don't forget general encyclopedias as well as specialist works.

Use the web: search for sources in specialist archives as well as using Google.

Make a selection of primary, secondary, and tertiary sources to give you different kinds of information: the author's actual words, detailed accounts and interpretations, or quick factual references.

Think carefully about the kinds of questions you want to ask, maybe under the headings "Where and when," "What and how," and "Who and why."

To think about

(1) Pick a historical mathematician (not Newton or Galois). Now find two primary sources, two secondary sources, and two tertiary sources about him or her. If you're using the web, try to find them at several different sites. If you picked someone unusual, you might have to look in the bibliographies of the

tertiary sources to track down the others; but see if you can do it, and think about what you learn along the way about searching for different kinds of sources. Which are the easiest to use? Which take us closest to the historical period itself? Which do you find the most interesting?

(2) You'll find three different versions of Galois' death in various sources—a duel, a political assassination, or a sort of political suicide. Find out about them—and decide which you think is most convincing and why.

(3) Look through either Fauvel and Gray's *The History of Mathematics: A Reader* or Victor Katz's *The Mathematics of Egypt, Mesopotamia, China, India, and Islam* and find a text with no known author. Now see what you can deduce from the text and the commentary about that unknown author.

(4) The following is part of a letter from Sophie Germain to Carl Friedrich Gauss, written in 1804.

> Sir, your *Disquisitiones arithmeticæ* have long been the object of my admiration and of my studies. The final chapter of that book contains, among other remarkable things, the beautiful theorem contained in the equation $4\dfrac{\left(x^n-1\right)}{x-1} = Y^2 \pm nZ^2$; I believe that it can be generalised to $4\dfrac{\left(x^{n^s}-1\right)}{x-1} = Y^2 \pm nZ^2$, n always being a prime number and s any number. I am attaching to my letter two proofs of this generalisation. After having found the first I sought how the method which you used in article 357 might be applied in the case which I had to consider. I

did this work with the more pleasure since it gave me the opportunity to familiarise myself with that method which, I have no doubt, will yet be in your hands the instrument of new discoveries. I have attached to this item certain other considerations. The last relates to Fermat's famous equation $x^n + y^n = z^n$, whose impossibility in whole numbers has not yet been proved except for $n = 3$ and $n = 4$.[8] I believe I have succeeded in proving that impossibility for $n = p - 1$, p being a prime number of the form $8K [+] 7$. I am taking the liberty of submitting these attempts to your judgment, being persuaded that you will not disdain to enlighten, with your opinions, an enthusiastic amateur in the science which you cultivate with such shining successes.

Nothing equals the impatience with which I look forward to what follows the book I have in my hands. I have informed myself that you are working on it at the moment; I will spare no effort to obtain a copy as soon as it appears. . . .

I feel that there is a kind of temerity in importuning a man of genius, when one has no other claim on his attention than an admiration necessarily shared by all his readers.

. . .

I did not want to tire your attention by multiplying the remarks of which your book has been the occasion

---

[8] Pierre de Fermat (1601–1665) famously claimed to have a proof that the equation $x^n + y^n = z^n$ possessed no solutions in positive whole numbers for n greater than 2. Fermat's proof being lost (and almost certainly flawed), "Fermat's last theorem" remained an unproved conjecture (until 1994).

for me. If I can hope that you will receive favourably those which I have had the honour to communicate to you, and that you will not find them entirely unworthy of a response, please address it to M. Silvestre de Sacy, member of the Institut national, rue Hautefeuille, Paris. Please believe how much I will value a word of opinion on your part, and accept my assurance of the deep respect of your very humble servant and most assiduous reader.

—Le Blanc

(There's a shorter extract in Fauvel and Gray's *Reader*, p. 496, and if you can read French, the original is on Gallica, in the *Oeuvres philosophiques de Sophie Germain*, pp. 254–58.)

What do you learn about Sophie Germain from this letter?

CHAPTER 3

Paper and Ink

I've got an eighteenth-century book in front of me. It's rather battered: a few of the pages are torn, and the binding has some ugly repairs. The paper is yellow, but the ink hasn't faded much, and the printed text is still completely readable. It's actually a more sturdy object than most modern paperbacks, and it'll probably last longer than a lot of modern hardbacks too.

Take a look at the pictures of it, in figures 3.1–3.3. As you can see, it's a one-volume encyclopedia or textbook, aimed at "young men." It's in six parts—one of them is on geography,

FIGURE 3.1: *The Young Man's Book of Knowledge* (exterior).

THE
YOUNG MAN's
Book of Knowledge:
BEING A
PROPER SUPPLEMENT
TO THE
YOUNG MAN's COMPANION.
IN SIX PARTS, *viz.*

PART I. Of Knowledge in general; the Advantages of gaining it early, with a Definition thereof. Of God, his Essence and Attributes. Of the Origin of Nature, and first Formation of Things. Of the Creation, Fall, and Restoration of Mankind.

PART II. Theology, containing an Account of the Religion and Laws of Nature. Supernatural Theology. Observations on the Holy Scriptures, which teach us the Knowledge of God and our Duty. Account of Judaism, Paganism, Mahometanism, and Christianity. Of the Sects of the Jews. Different Tenets of the principal Sects or Professors of Christianity. Of the Heathen Mythology, and Alphabetical Account of the Heathen Deities.

PART III. Natural Philosophy in general.

PART IV. Geography, in a Manner entirely new: Containing, (by Question and Answer)

1. A general Description of the four Quarters of the World. 2. The Situation, Extent, and chief Cities of the several Kingdoms and Countries of each Quarter. 3. The Nature and Description of the Globes, and Explanation of the Terms used in Geography. 4. Tables of the Latitude and Longitude of several principal Places, with many useful and necessary Problems on the Terrestrial and Celestial Globes.

PART V. Geometry and Astronomy, Navigation, and Plain-Sailing; with many useful, easy, and instructive Problems for the young Practitioner in the further Knowledge of those Sciences.

PART VI. Of Music and Vibration. Definition of Music. Gamut or Scale, and Explanation of dividing Notes in Time, &c. Of the Diatonic Scale, an Explanation. Different Keys, Time, Bass, &c. &c.

By D. FENNING,
Author of the ROYAL ENGLISH DICTIONARY, UNIVERSAL SPELLING-BOOK, USE of the GLOBES, &c. &c. &c.

THE FOURTH EDITION,
Revised, corrected, and greatly improved.

The GEOGRAPHICAL, GEOMETRICAL, and ASTRONOMICAL Parts,
By Mr. MOON, of SALISBURY.
The MUSICAL Part,
By Dr. ARNOLD.
And the other Parts by a CLERGYMAN of the CHURCH of ENGLAND.

LONDON:
Printed for S. CROWDER, in Pater-noster-Row, and B. C. COLLINS, in SALISBURY. 1786.
PRICE THREE SHILLINGS, BOUND.

FIGURE 3.2: *The Young Man's Book of Knowledge*, title page.

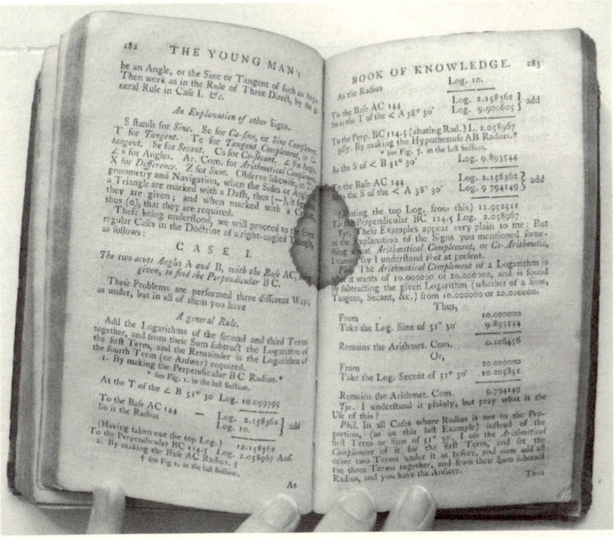

FIGURE 3.3: *The Young Man's Book of Knowledge*, pp. 282–83.

with some exercises in spherical trigonometry, and another is on geometry, astronomy, and navigation. These two sections give lots of mathematical information—so they tell us a lot about what a young man in 1786 might have been expected to know.

At least two of the book's previous owners have written in it, sometimes in ink and sometimes in pencil. These annotations—they cover the inside front and back covers almost completely—have faded more than the printed text, but they're still easily readable. There are only a few notes among the actual text of the book—there's no rough working of the exercises, or anything like that—but there are a few interesting jottings, which I'll come back to.

If you ever get the chance to handle an old book, take it. Books are real physical objects, not just "disembodied texts"—

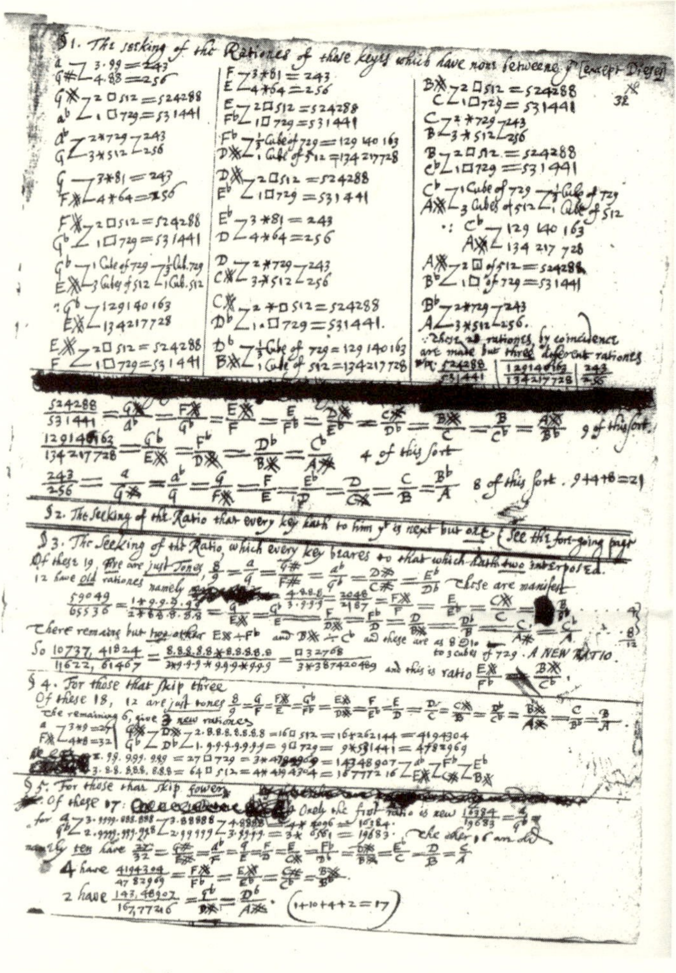

FIGURE 3.4: John Pell, page of musical calculations.

and in this chapter you'll learn some of the ways that that can help you find out about historical mathematics. We'll mostly consider published books, but what you'll learn will also apply to handwritten manuscripts and all the other sorts of objects where historical mathematics is found.

There's a lot you can learn from a physical book that you wouldn't get from the same text reprinted in a modern book or displayed onscreen. Look again at figure 3.3 and compare it with figure 3.4, a page of mathematical working by John Pell, an English mathematician who lived around the same time as Isaac Newton. In a disembodied text the two might look similar. Here they look very different. What are the differences, do you think, and what do you learn from them about the history of mathematics?

Pause for thought

How to find out

This is all very well, of course, but who actually has access to very old books? Even if you're at an institution with a big library, it probably doesn't have that many historical mathematics books, and it probably won't let you touch them if it does.

What can you do instead? If you're presented with a disembodied text—like the ones in chapters 1 and 2—and you want to find out about the physical book it came from, what can you do?

The obvious thing is to hunt around for pictures of it. If you search on the web for images of the two main extracts in chapters 1 and 2, you'll do quite well. Page images of Newton's writings are pretty easy to find. If you're at an institution which subscribes to Early English Books Online or Eighteenth Century Collections Online, you'll have access to page images of a great many English books from before 1800—though they don't give you any impression of things like the book's size and weight, the thickness and the feel of the paper, and so on; they're almost halfway to the disembodied texts you find in a modern edition. What about pictures of the book "from the outside," so to speak? They can show you what the printed pages really look like, as well as how big the book is and things like what sort of binding it has. Again, there are plenty for Newton—an image search for "Newton Principia" turns up quite a few. Galois' last letter is a bit harder to track down, but the published version is on Gallica, and www2.maths.ox.ac.uk/arg/misc/galois.shtml has images of that and some of his other manuscripts, which are interesting to see. (I found it by image searching for "Galois manuscript.")

If it's even slightly famous, there'll be pictures of it in odd places on the web. (I just image-searched for "Euler Introductio analysin" and found a few dozen page images from the book— Leonhard Euler's *Introductio in analysin infinitorum* (1748)— as well as one of the book from the outside, showing a hefty textbook-sized volume.) If it's in a major library or museum, there's a chance that the library will have pictures of it on its website—for example, Cambridge University Library has pictures of Blaise Pascal's treatise on the "arithmetic triangle." For more obscure things you might be lucky—the websites of antiquarian book dealers or auctioneers also often have images of the books they handle, for example. And the book listings on ebay usually include a random selection of old books about mathematics, and they almost always include pictures of the book from the inside and the outside.

What if you just can't find a picture of the book you're interested in? Well, there's still hope. Instead of a picture, you may be able to find a description of the book, which can help more than you might think. Try looking it up in the catalog of a large library. Looking up "Euler analysin" in the catalog of the Library of Congress finds the book mentioned above, Euler's *Introductio*, and choosing "full record" will give you, among other things, the following information:

Published/created: Lausannae, M. M. Bousquet, 1748.
Description: 2 v. front., diagrs. 28 cm.

Rather cryptic, but a bit more detective work will tell you that Lausannae is a Latin form of the place-name Lausanne, in Switzerland (there's more than one glossary of Latin place-names on the web, and they can be very useful on occasion). M. M. Bousquet is the name of the publisher. The physical description here is quite short; expanding the abbreviations gives "2 volumes, frontispiece, diagrams, 28 cm." That is, the book is bound in two volumes (so it's a long book), it has a frontispiece and diagrams, and one of its dimensions (presumably the longest) is 28 centimeters. That's fairly large, and together with the fact that the book is bound in two volumes, tells us that this is a hefty work: perhaps the sort of thing you'd be more likely to find in a library than at home, and certainly not the sort of thing you can carry around with you to work.

Web searching for images and searching library catalogs for descriptions are probably the two best ways to find out physical information about an old book. But there are sometimes other sources of information too. For example, sometimes a modern

How to Find Out

If you want to know what an old book was physically like, you can . . .

Find page images of it on the web.

Find pictures of it from the outside—try library sites if a general search doesn't work, or booksellers.

Find a description of it in a library catalog.

Find descriptions or pictures of it in a modern edition of the book.

Or if you're really lucky, you might be able to see a copy of the book itself in a library or belonging to—or borrowed by—one of your teachers.

edition of the book contains some pictures of its older counterpart—or sometimes it has a physical description of the original book. One way or another, you can often find out quite a lot about an old book.

What does it mean?

Back to *The Young Man's Book of Knowledge* and the page from John Pell. I asked you to think about the differences between figures 3.3 and 3.4 and what you can learn from them about the history of mathematics. Here are some ideas.

The manuscript is rough work; it has survived to the present only by accident; in the normal course of things it would probably have been thrown away once its purpose had been served. The book is different—it was meant to be a fairly long-lasting, durable object: something you could come back to and refer to again and again. The manuscript represents a quick, cheap way for Pell to record his mathematical thoughts—and we can see from the way he crammed the paper full of notation that he didn't want to waste any space. Compare that with the quite large areas of blank paper in the book. Though it was quite a cheap book, it's still a kind of mathematical writing where clarity sometimes matters more than space.

The manuscript is one of a kind. If you want just one or two copies of your mathematics, writing it out by hand is the cheapest and easiest way to do it. But if you want a few hundred copies, or a few thousand, a printed book is much cheaper and easier. I don't know how many copies of this book were printed, but since it got as far as a fourth edition (see figure 3.2, near the bottom), it was presumably selling quite well. There might easily have been thousands, or even tens of thousands, of copies of it around.

The manuscript is "private" mathematics, so to speak—it might have been shown to a few friends or it might not have been shown to anyone else. You can tell that from how chaotic it is and the fact that it's unclear what it means—though presumably it was clear to Pell. But the book was meant to be available to anyone who might be interested—there were lots of copies of it, and they were reasonably cheap (it's hard to say what else three 1786 shillings might have bought, but it's not a lot of money), and the book makes it quite clear that it's aimed at young men pretty generally. Plus, it's quite a small book—it would fit in your pocket. It is meant for people to carry about with them, maybe to school or to work—not to be kept in a library. So, if the manuscript is a prestigious, exclusive sort of object, the book is the opposite: low-prestige and widely available.

Those are a few things you can learn by comparing figures 3.3 and 3.4. Lots of kinds of objects can be sources for the history of mathematics—books, manuscripts, carved inscriptions, other media like the bamboo and silk used in classical China, the clay tablets from ancient Mesopotamia, etc. They all have their own peculiarities, but you can learn a lot from any physical object by asking the same general types of questions we asked here.

First: Is it throwaway or long-lasting? Is it a *durable* object or not? Some kinds of writing are consciously throwaway, like pencil jottings on cheap paper. Others are intended to be long-lasting or permanent records—the most obvious example would be an inscription carved on stone. But beware—the solidity of a physical object can sometimes be misleading. Some of the oldest mathematical texts that exist are writings incised on clay tablets from ancient Mesopotamia. Mostly the tablets would have been dipped in water to remove the writing and then reused. The ones which survive come from relatively unusual situations when the tablets were lost, abandoned, or reused as building materials. Over the centuries they've dried, or been baked, and hardened, so that what were originally soft, easy-to-recycle objects are now solid, though fragile.

A text meant for throwing away could be a piece of rough working—if it turns out to be correct and interesting, it might get written up in a more durable form. It could also be an exercise done by someone learning mathematics—as in a school exercise book. A text that is meant to be kept could be a work of reference or a private memory aid. It could record new results that the author thinks are significant, or it could just be a new presentation of existing mathematics, like a long-lived textbook.

Second: Is it *prestigious*? How much money or effort has gone into it? Is it a messy jotting for a friend or an expensive gift for a prince? This isn't quite the same as durability—though prestigious items quite often do last a long time, one way or another. A book can be printed on large, thick paper and bound in expensive leather—or it can be a cheap modern paperback. A carved inscription can be someone graffiti-ing on the wall of a cave—or an elaborate decoration on a huge monolith.

Third: Is it for everyone, or not? How widely was it *disseminated*? A book might be printed in thousands of copies—a carved inscription might be seen by everyone who walks by. An exercise written on a school slate might never be seen by anyone except its writer—a typed report of a new discovery might be thrown away as soon as the author spots an error in the proof. Some authors write only for themselves—others make huge efforts to make sure a large number of students will understand what they have to say.

This is different from durability or prestige, but it's related to them—something that's very widely available doesn't often have much prestige value, for instance. If you mass-produce a book, you might not be very concerned to make sure the copies last hundreds of years.

Books often give clues about how widely they were meant to be disseminated. But you can also sometimes find out, by searching in national library catalogs (like COPAC or the Library of Congress),[1] just where the surviving copies are—or at any rate some of them. That can give you a sense of how widely the book was in fact disseminated, though like all of these kinds of information you need to use it with caution.

There's a slight problem here, and it's one that comes up all the time in history. A cheap textbook might be carried around in someone's bag or pocket and be used and used until it falls apart, then thrown away. An expensive reference book might be used just as much, but it would be kept at home or in a library and be treated more carefully—so it would be more likely to survive until the present day. On the other hand, a cheap textbook might have been originally printed in many thousands of

---

What Does It Mean?

If you're studying a physical object from the past you can ask . . .

Is this a throwaway object or a long-lasting one? (durability)

Has a lot of money or effort gone into it? (prestige)

Is it for everyone, or not? (dissemination)

And, why has it survived? Is it normal or unusual?

---

[1] copac.ac.uk, catalog.loc.gov.

copies, making it more likely that some of them would survive to the present. The situation is different for different books, and hard to predict. What that can mean is that the sources we have to work with don't give a very good representation of the past—they might even leave out some of the things that were most common at the time. The objects that do survive sometimes do so just because they're unusual, expensive, or hard to destroy. Sometimes, in fact, you can learn a lot by asking, Why has this survived?

The box summarizes the kinds of questions we've mentioned: questions you can ask when you want to study a physical object from the past. Next we will think about how these objects actually get to us, and what can happen to them along the way.

How it got to us

More people than just the author leave their mark on a book or a manuscript from the past. Look again at figures 3.1–3.3 and figure 3.5. Who are the people who have left their marks on this book? How many of them can you name? What can you find out about them?

Pause for thought

You probably thought of quite a few people. Here are some of my ideas about them.

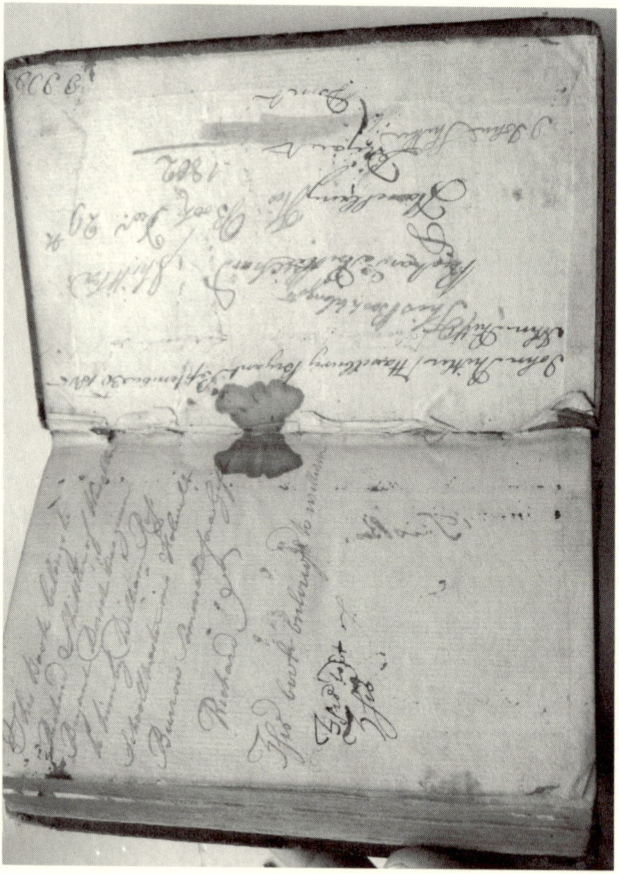

FIGURE 3.5: *The Young Man's Book of Knowledge*, annotations.

On the title page there are two names other than that of the author. It says the book was "printed for" S. Crowder and B. C. Collins. What does that mean? Probably that they paid for the printing, sold the book, or both. If you look up Collins in the *Dictionary of National Biography*, you'll find a Benjamin Collins—he's described as a "newspaper proprietor and publisher," but he died in 1785. His son, Benjamin Charles Collins, is probably the person who was involved in our book. If you look up Crowder, you'll only find a John Crowder, "printer and mayor of London," alive at the right time—maybe S. Crowder was a relative.

If Crowder was the printer and Collins was the publisher, what does that mean? Well, it probably means Crowder—or his employees—made a lot of the detailed decisions about how the text and diagrams were laid out on the page, and he employed the people who actually did the physical job of printing the book. (Sometimes the mathematical notation is chosen or altered by the printer—or the author uses notation that he knows will be feasible to print. Have a look at the matrix notation shown in Stedall's *Mathematics Emerging*, pp. 589–92.) He might have risked some of his own money in the venture, or he might not have. Collins, as publisher, might have put up most of the money for printing the book and taken a hefty share of the profits. Publisher, author, and printer would have made decisions about what kind of paper to use, how big each page should be, what size of print should be used, and therefore how many pages the book would have and how heavy it would be. They would have decided how many copies to print and how much they would be sold for.

Another interesting way to find out about Collins and Crowder is to search for their names on the title pages of other books from the same time. If you have access to Eighteenth-Century Collections Online, you can do that on the "advanced search" page. It'll give you an idea of what sort of books they were involved with—and just how many books as well.

Once it was sold, other people had the chance to make their mark on the book. In figure 3.5 you can see that some of them wrote their names on the front and back covers. These kinds of annotations on covers and in margins can be the most fascinating things about old books (you might have heard about Fermat . . .).

In this case we can trace how the book was passed from owner to owner. Annotations, found throughout the book, say:

Wilhelmus Ross His Book, Anno, Domini 1792 . . . cost
    3 Shillings + six pence
This book belongeth to william
This Book belongs to Richard Shittler of Haselbury
    Bryant Dorset was given to him by William Ross
    Schoolmaster in Holwell Burrow Somerset for a Gift
Richard Shittler his Book gave to him by Wm Ross Holwell
    Somerset 1792
This Book belongs to Richard Shittler His Book Dec. 29th
    1802
Richard Shittler his hand & pen Amen October 25 1803
John Shittler of Haselbury Bryant September 30 1805
Richard Shittler's Book December 3 1806

Inside the book, on the back of an illustration of the solar system, there's this, in pencil:

[*illegible*] saild from Poole May 4—1811—for Spithead to
join Convoy for [*illegible*] Land
May 22 in the Latitude 44 North

Maybe this doesn't tell us much about the mathematics in the book. But it does tell us about the book's history, and that's part of the history of mathematics as much as anything is. It was sold 10 years after it was printed, either by a bookseller or by a previous owner, to a schoolmaster—William Ross. (He bought it for more than its original price of three shillings, so possibly this book had become a minor classic by then—or maybe it was just the result of inflation.) Soon afterward he gave it to Richard Shittler—who I guess was one of his pupils and who wrote his name in the book on three different occasions. (Was he just doodling in idle moments, or can you think of some other reason why he would do this?) Then John Shittler—Richard's younger brother, perhaps?—had the book for a while around 1805. The part about sailing for Spithead could mean two things—the book's owner either had a relative at sea or was at sea himself. I guess it was the latter—this young man could have been in the Navy during the Napoleonic wars, and he might have used this book to help him learn the mathematics of navigation. There's nothing to tell us about where the book went afterward—I bought it through ebay from a dealer in Berkshire.

By thinking about *who* has marked this book, we've also found out something about *where* it has been—printed in Salisbury, bought and taken to Somerset, given away and taken to Dorset, and then taken onto a ship where it sailed from Poole (on the south coast of England) to Spithead (in the English

How It Got To Us

If you're looking at an object from the past, you can ask . . .

Who are the people who've left their mark on it? Think about printers, publishers, copyists, readers, etc.

Where has it been? Does the place of publication or any annotations tell you anything interesting?

What has it been used for? Were the people who owned it teachers, students, workmen, etc. Were they reading it for pleasure, for professional advancement, because a teacher made them, etc.?

Channel) and beyond (44 degrees north is roughly the latitude of Bordeaux or Nova Scotia). We've also guessed about *what* it was used for—when it was taken to sea, it probably helped its owner learn navigation.

Transcription

There's one more thing I'd like to think about in this chapter, and that's what happens when these objects—books or whatever else—get turned into disembodied texts. From what we've considered already in this chapter, you'll have an idea of the sort of information that can get lost. Other things can happen too—this is another way that someone other than the original author makes a mark on a piece of historical mathematics.

Look again at figures 3.3 and 3.4: a page from the *Young Man's Book of Knowledge* and a page from John Pell. Choose one, and imagine making a disembodied text version of it. What would you have to do? How would your version be different from the original? Would it say more or less?

Pause for thought

If you chose the manuscript, the first thing you'd have to do would be to decipher the handwriting. In some places it's quite clear, but there are places where it isn't. Would you mark them in some way in your version? And what about the crossings-out? Sometimes you can figure out what word has been crossed out, but sometimes you can't—would you include them in the disembodied version, maybe in square brackets, or leave them out? And these two sources are in relatively good condition, but a lot of books and manuscripts are much the worse for wear. What would you do if there were parts you just couldn't read at all?

If you chose the printed text, you wouldn't have most of these problems—but you would have some others. The layout on the page is quite important here—would you be able to copy it exactly? Would it matter if you couldn't? What about the footnotes? Would you put them at the bottom of your page or leave them where they are in the original—or might you want to put them somewhere else, or leave them out altogether since they refer to pages that you're not transcribing? Then again, what would you do with the angle sign? Would you be able to use it in your version, or not?

With the manuscript you also have the problem of how things are arranged on the page. You could try to mimic it, or you could set things out in what seems to be their logical order. And what if the author inserts extra ideas as afterthoughts? Should they be in footnotes, or be inserted into the main text?

What about the spelling ("hypothenuse" in the book, for example)? Would you modernize it or leave it alone? Or the abbreviations—"Arithmet. Com.," for example? You could expand it to "Arithmetical Complement" or "Arithmet[ical] Com[plement]" or leave it alone.

What if you found a mistake in the original text? You could correct it, or leave it, or insert a footnote pointing it out. But which is best?

If there were diagrams, or more complicated algebraic notation, you'd have to make decisions about those too. You'd probably end up using modern algebraic notation, even if it was a bit different from what was in the original source—and you'd probably end up redrawing the diagram, which might change it a bit (see exercise 2 at the end of this chapter).

Finally, you'd probably want to give your version a title. If you had the book in front of you, you could find out that pages 282–83 are in a section called "Practical Trigonometry, *or to measure or find the Sides or Angles of any Triangle*." For Pell you might use the caption I've given to figure 3.4: "John Pell, page of musical calculations"—but Pell didn't give this a title himself. But even so, a disembodied version of just one or two pages of text would have lost a lot of its context: Why are we doing these calculations? What comes before? What will we do next?

Transcription

If you're asked to produce a disembodied version of a historical source, you can . . .

Clean up any crossings-out or illegible parts, or not.

Change the layout, or not.

Modernize the notation, the spelling, and the abbreviations, or not.

Redraw the diagrams, or use facsimiles of them.

Give it a title, or not.

But if you're looking at someone else's disembodied version, you can remember . . .

The original might have had crossings-out or words that were impossible to read.

It might have been laid out very differently.

It might have had different notation, spelling, and abbreviations.

Errors might have crept in.

The diagrams might have been redrawn.

It might have had a different title, or no title.

It might have had a context that helps to interpret it.

So if you can have a look at the original, do!

You can see that there are a lot of decisions to make. An editor who does this kind of thing for real has to make these sorts of choices all the time—you can be grateful that so much work has been done for you, but you should also remember that this is another person who has left a mark on the text. When you are faced with a disembodied text, if you can have a look at the original, do so—you might be surprised at how different it is. The box summarizes the different things that editors can do to a piece of mathematics.

Conclusion

You've learned a lot in this chapter about how historical mathematics is embedded in particular objects. You've learned how to find out about those objects and what you can learn from them—and you've also learned some of the ways a disembodied text might be different from the original. The box summarizes all this.

In chapter 2 we thought about the *writers* of historical mathematics. In chapter 4, we will think about the *readers* they were writing for.

To think about

(1) Find a copy of *The Mathematics of Egypt, Mesopotamia, China, India, and Islam*, edited by Victor Katz, and choose one of the pictures of mathematical sources. Even though you can't read

---

Paper and Ink

Every piece of historical mathematics comes from some kind of physical object. You can . . .

Find out what sort of object it was: a book, a newspaper, a handwritten manuscript, an inscription carved on stone, a clay tablet, etc.

Find out something about that particular object: look for pictures or descriptions on the web, in library catalogs, or in another edition of the text.

Think about the object's durability, its prestige, and its dissemination: What does the object tell you about them?

Find out how it got to us: who has left a mark on it, where it has been, and how it was used.

Compare a modern edition of the text with the original object or a picture of it, and try to understand why they are different.

---

it, how might you apply the questions we have asked in this chapter?

(2) Take a look at the diagram—based on one by Fermat—on p. 362 of Fauvel and Gray's *The History of Mathematics: A Reader*, and the one on p. 483 of Katz's *A History of Mathematics: An Introduction* (second edition). How many differences can you find? What do you think the reasons for them are? Fermat's original diagram doesn't exist any more, but there's

an older version on the web at http://www.maths.uwa.edu.au/ ~schultz/3M3/L14Fermattangent.html. What does that tell you?

(3) Find out the age of the oldest existing manuscript of Euclid's *Elements*. What consequences does that have?

(4) Actually produce, using TeX or a word processor, a printed version of a picture of a mathematical text of your choice. What decisions do you have to make? Do you think your version is satisfactory? Why?

(5) Go to the website of a science museum or history of science museum (try the one in Oxford, or the Musée des Arts et Métiers if you read French)[2] and find a picture of a mathematical instrument, preferably with a description of it. How can you apply the questions we've asked in this chapter? See what you can find out to answer them.

[2] www.mhs.ox.ac.uk, www.arts-et-metiers.net.

CHAPTER 4

Readers

1. *Definition I Variable quantities are those that continually increase or decrease; and constant or standing quantities, are those that continue the same while others vary.*

As the ordinates and abscisses of a parabola are variable quantities, but the parameter is a constant or standing quantity.[1]

Definition II The infinitely small part whereby a variable quantity is continually increased or decreased, is called the differential of that quantity.

For example: let there be any curve line *AMB* whose axis or diameter is the line *AC*, and let the right line *PM* be an ordinate, and the right line *pm* another infinitely near to the former.

Now if you draw the right line *MR* parallel to *AC*, and the chords *AM, Am*; and about the centre *A* with the distance *AM*, you describe the small circular arch *MS*: then shall *Pp* be the differential of *PA*; *Rm* the differential of *PM*; *Sm* the differential of *AM*; and *Mm* the differential of the arch *AM*. In like manner, the little triangle *MAm*,

---

[1] Briefly, a parabola is defined as the set of points equidistant from a given point (the focus) and a given line (the directrix); we call the vertex the top or middle point of the parabola, and its line of symmetry, its axis. For any point *P* on the parabola, the ordinate is the perpendicular distance to the axis, and the abscissa is the distance along the axis from the foot of that perpendicular to the vertex; obviously these are variables, depending on the choice of *P*. The width of the parabola at the focus is called the parameter, a constant.

whose base is the arch *Mm*, shall be the differential of the segment *AM*; and the small space *MPpm* will be the differential of the space contained under the right lines *AP*, *PM*, and the arch *AM*.

Corollary It is manifest, that the differential of a constant quantity (which is always one of the initial letters *a*, *b*, *c*, etc. of the alphabet) is 0: or (which is all one) that constant quantities have no differentials.

Scholium The differential of a variable quantity is expressed by the note or characteristic *d*, and to avoid confusion this note *d* will have no other use in the sequence of this calculus. And if you call the variable quantities *AP*, *x*; *PM*, *y*; *AM*, *z*; the arch *AM*, *u*; the mixtlined space *APM*, *s*; and the segment *AM*, *t*: then will *dx* express the value of *Pp*, *dy* the value of *Rm*, *dz* the value of *Sm*, *du* the value of the small arch *Mm*, *ds* the value of the little space *MPpm*, and *dt* the Value of the small mixtlined triangle *MAm*.

2. *Postulate I* Grant that two quantities, whose difference is an infinitely small quantity, may be taken (or used) indifferently for each other: or (which is the same thing) that a quantity, which is increased or decreased only by an infinitely small quantity, may be considered as remaining the same.

For example: grant that *Ap* may be taken for *AP*; *pm* for *PM*; the space *Apm* for *APM*; the small space *MPpm* for the small rectangle *MPpR*; the small sector *AMS* for the small triangle *AMm*; the angle *pAm* for the angle *PAM*, etc.

3. *Postulate II* Grant that a curve line may be considered as the assemblage of an infinite number of infinitely small right lines: or (which is the same thing) as a polygon of an infinite number of sides, each of an infinitely small length, which determine the curvature of the line by the angles they make with each other.

For example: grant that the part *Mm* of the curve, and the circular arch *MS*, may be considered as straight lines, on account of their being infinitely small, so that the little triangle *mSM* may be looked upon as a right-lined triangle.

4. *Proposition I To find the differentials of simple quantities connected together with the signs + and −.*

It is required to find the differentials of $a + x + y - z$. If you suppose x to increase by an infinitely small part, viz. till it becomes $x + dx$; then will y become $y + dy$; and z, $z + dz$: and the constant quantity a will still be the same a. So that the given quantity $a + x + y - z$ will become $a + x + dx + y + dy - z - dz$; and the differential of it (which will be had in taking it from this last expression) will be $dx + dy - dz$; and so of others. From whence we have the following.

Rule I For finding the differentials of simple quantities connected together with the signs + and −.

Find the differential of each term of the quantity proposed; which connected together by the same respective signs will give another quantity, which will be the differential of that given.

5. *Proposition II To find the differentials of the products of several quantities multiplied, or drawn into each other.*

The differential of xy is $ydx + xdy$: for y becomes $y + dy$, when x becomes $x + dx$; and therefore xy then becomes $xy + ydx + xdy + dxdy$. Which is the product of $x + dx$ into $y + dy$, and the differential thereof will be $ydx + xdy + dxdy$, that is, $ydx + xdy$: because $dxdy$ is a quantity infinitely small, in respect of the other terms ydx and xdy: For if, for example, you divide ydx and $dxdy$ by dx, we shall have the quotients y and dy, the latter of which is infinitely less than the former.

Whence it follows, that the differential of the product of two quantities, is equal to the product of the differential

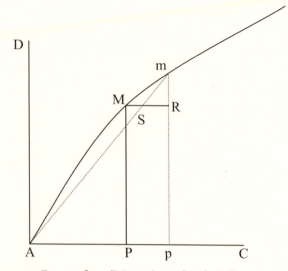

FIGURE 4.1: Figure 1 from l'Hôpital, *Analyse des Infiniment Petits.*

of the first of those quantities into the second plus the product of the differential of the second into the first.

—G. de l'Hôpital, *Analyse des Infiniment Petits, pour l'Intelligence des Lignes Courbes,* translated in D. J. Struik, *A Source Book in Mathematics, 1200–1800* (Harvard University Press, 1969), pp. 313–15.

The long quote I've given above was first published in 1696. The translation is based on one that first appeared in 1730.[2] To understand it, you can use the skills you've already learned in this book—translating the mathematics into modern terms, finding out something about Guillaume de l'Hôpital, and finding out what the book physically looked like that it was printed in. It's historically interesting for quite a few reasons.

[2] E. Stone, *The Method of Fluxions Both Direct and Inverse* (London, 1730).

What about the people who read the book? We've left them out so far, and now it's time to think about them. Look again at the quote above and see what you can work out about the people who read the book—or who were supposed to.

Pause for thought

The reader

I wonder what you came up with. As we did in chapter 2, let's call the reader X and see what we can work out about him or her. You can work out quite a lot about X from the way the mathematics itself is set out and presented here, and we'll come to that in a moment. First, let's look at the title. The original title was French, so the book was originally in French. Then, 34 years later, it was translated into English. Two things might come to mind about that. First, why not Latin—either for the original version or the translation? We saw in chapter 1 that Newton, in 1687, wrote in Latin so that his book would reach an international audience. In fact, his *Opticks* was originally published in English and then translated into Latin so that people on the Continent could read it. So what was going on here? The natural answer is that presumably X couldn't read Latin very well—maybe X was a student.

If it was worthwhile to translate it 34 years later, this was probably quite a popular book—but 34 years is a long time in mathematics. Something that was new and important in 1696 wouldn't have been cutting-edge in 1730. Maybe this is not new mathematical work but material for students: maybe X was a student—someone who wasn't doing new research in mathematics but who wanted to learn the basics of recent mathematics.

Now let's look at the contents of the text. It certainly doesn't read like cutting-edge research, even allowing for the fact that calculus was quite new at the time. X apparently knows absolutely nothing about calculus and even needs a definition of "variable quantities." X is guided very slowly through the most basic steps of defining and using differentials, with a diagram to help, and by the end of this fairly long extract X has reached the product rule for differentials. There are no steps of the argument that the reader has to fill in. X is an absolute beginner—someone with no experience at all in working through these ideas.

X has plenty of reassuring signposts—"Definition I," "Corollary," "Scholium," "Postulate I," "Proposition I," "Rule I," etc. These things make it clear where you are at any point: they reinforce the structure of the explanation. They also remind X of Euclid's *Elements*, a work probably studied at school—in fact it's so like Euclid that it almost comes across as a caricature of "proper" mathematics writing. Comforting for X, perhaps, for whom the book contains new and difficult ideas.

Can we go into more detail? I think we can. The reader isn't completely ignorant of mathematics—X does have some knowledge and skills that l'Hôpital can rely on, even if they're pretty basic. X can deal with the use of letters to stand for variables or constants, for example, and is familiar with basic algebraic notation—plus and minus signs and the fact that xy means x times y. The reader knows how to multiply out $(x + dx)(y + dy)$.

X also knows some geometry. The very first definition has a casual reference to three terms—the ordinates, abscissae, and parameter of a parabola—which most people today wouldn't have heard of, yet l'Hôpital doesn't bother to explain them. And X can understand a geometric diagram with various labeled points.

Finally, X is actually reading the book. Not everyone reads books about calculus. L'Hôpital, at least in this extract, doesn't give any reasons why you'd want to know these things—he doesn't say what they're useful for, or point out that they're new

---

The Reader

If you want to know who was supposed to read a piece of historical mathematics, you can . . .

Think about the *continuity* of the argument. Does the reader need everything explained step by step, or can some gaps be left for the reader to fill in?

Think about the *content* of the piece. What does the reader need explained—and what does the reader know already?

Think about the *tone of voice* the author uses. Does the reader have to be reassured? Is the reader likely to be critical? What sort of person would you speak to or write to in this way?

Think about the reader's *motivation*. Why read this? Does the writer make the mathematics sound exciting, interesting, useful, etc.?

---

and exciting. He assumes X is interested enough to keep reading anyway.

So who is X? A person with some mathematical training but not a huge amount. Someone who probably knows more geometry than we do, but for whom this book is probably a first taste of calculus. And—since X is reading a book about calculus—someone who is either being made to read this book or is interested in the subject.

So X is probably a student—or possibly an educated person who's interested in mathematics.

When you read a piece of historical mathematics, you can often build up a picture of the reader like this. You can think about things like the structure and presentation, the kind of mathematics and whether it was old or new, and the assumptions the author makes. You can figure out why it's written the way it is. The box summarizes the questions you can ask as you read.

Mathematical genres

One more thing, before we think some more about the readers of a piece of mathematics. L'Hôpital's book is a textbook (in fact, it was the first textbook about the calculus). That's one example of a *genre* of writing—a type of writing with its own particular assumptions and ways of doing things. Other genres of mathematical writing are letters, research papers, popular books and articles—even lecture notes. (Can you think of some others?) This can be a handy way to think about what sort of writing you're looking at and why it might have been written.

A letter is written for a particular person (at least, it usually is). A research paper is written for other specialists in one particular branch of mathematics. The notes you take in a lecture are probably written just for you. The different kinds of readers often account for the different styles of presentation you'll find in these different kinds of writing.

Sometimes you'll know that you're looking at something from a letter, a textbook, and so on. The title might give it away, or there might be a preface in which the author says who the piece is for. That can help you too, because it means you have some clues about the kinds of things to look for. In a letter, for example—think of Galois' letter in chapter 2—you'd expect to find new ideas, maybe things the writer isn't quite sure about yet, expressed pretty briefly. The expected reader is probably another mathematician, and is prepared to put in some effort to understand the letter. In a piece of popular writing, where the expected reader could be any interested person, you'd expect to find a few mathematical ideas explained quite carefully, maybe without using much notation. There'd be explanations of why this is interesting or useful—why the reader should keep reading.

If you think about the other kinds of mathematical writing, you can come up with some ideas about what you might expect to find in them. So this can work both ways—thinking about *genre* can help you know what to expect, or it can help you make sense of what you find. The box summarizes some useful questions to ask.

Mathematical Genres

If you want to figure out what *kind* of mathematical writing this is, you can ask . . .

Is the mathematics new or old?

Is this a text which tries out new ideas, or does the writer seem confident of being right?

Does the writer expect a lot of knowledge and effort from the reader?

Does the writer expect the reader to have any other help (a teacher, colleagues, etc.) with understanding this?

Does it seem similar to any of the kinds of mathematical writing you've come across before—textbooks, research papers, popular books, letters, dictionaries, newspaper articles, private notes, etc.?

Readers or no readers

Look again at Figure 3.4 in the last chapter. Who was meant to read this? How can you tell?

Pause for thought

I admit it—it was a trick question. Probably no one was meant to read it except Pell himself. You can tell because of how dense and cryptic it is—presumably Pell understood what was going on here, but he's done little to make it comprehensible to anyone else. There aren't many words to explain things. You can just about figure out what he's doing, but it takes a while. If he'd ever given this mathematics to someone else it would have needed rather more explanation with it.

When you think about it, a lot of historical mathematics must be like this—rough jottings, first drafts, notes on the back of something else that were never meant to be read by anyone else, at least until they'd been tidied up. When notes of that kind do end up being read by other people they can be very confusing. A good example is the eight volumes of Newton's *Mathematical Papers*. They're writings that, mostly, Newton didn't imagine anyone else would ever read. Like our page from Pell, a lot of them are complicated and chaotic—and in this edition they come with a huge amount of commentary from the editor, to make them comprehensible at all.

(There are other ways mathematics can be private. Sometimes it's written in code so the author can claim priority later on without giving away what the discovery is. Sometimes the author actually tries to destroy it, maybe because there's a mistake in the proof. Or sometimes it never gets written down at all, but stays in the author's head. That doesn't always stop it from being influential and important—think of Fermat's "proof" of his last theorem that sparked off 350 years of mathematical research despite being (1) lost and (2) almost certainly wrong.)

This kind of thing can have an important place in the history of mathematics. In a way, private writings are another genre of mathematical writing—just like the more public genres we considered in the previous section. Can you think of some reasons we might want to read private mathematics—things we can learn from it that we can't learn anywhere else?

The most obvious reason is one I mentioned in chapter 1— private mathematics can help us to see how historical mathematicians *thought*. They give us a glimpse of their mathematics in its first state, before it was polished up for sharing with others.

Following on from that, private mathematics can help us to understand public mathematics. In Newton's *Mathematical Pa-*

Readers or No Readers

If you're looking at a piece of private mathematics, you can . . .

Check whether the mathematics is correct. Was this a method that didn't work, a proof that contained a mistake?

Compare it with a published version of the same thing, if there is one. How did the writer's ideas change? How did things get polished before publication?

See whether the writer is being deliberately secretive. Was there something to gain by keeping this mathematics quiet?

Get a glimpse of how the writer *thinks*. Does the text use words, symbols, pictures, etc.? Are there gaps in the mathematics, or surprising insights?

pers there are drafts of some parts of the *Principia*, for instance. They can help us understand what Newton meant in the published version of the *Principia*—perhaps showing us how he arrived at some of the ideas there, or why he thought of putting things the way he did.

And private mathematics also gives us a window onto all the mathematics that never got published. We can see mathematicians struggling with problems they *can't* solve, trying out methods of proof that *don't* work, and so on. That's part of the history of mathematics just as much as the successes, the things that *do* work. It's a part that we don't often get to see—so if you do get a chance to study some private mathematical writings, take advantage of it.

Who actually read it?

There's one more postscript to this chapter. So far we've looked at who was *supposed* to read a particular piece of mathematical writing—what sort of readers (if any) the author expected. But who *actually* read it?

Most of the time we just don't know. How could you possibly figure out who actually bought copies of a particular book, or saw a particular manuscript or letter?

The example we looked at in chapter 3 shows some of the ways you can get clues about this. The title page says it's the fourth edition—so three editions had already sold out, and the publisher must have thought a fourth one would sell as well. The book must have had quite a few readers. And the annotations we looked at tell us who some of them were—teachers, students, and possibly someone at sea.

That's a rare opportunity, though. When you do have this sort of information, it can be interesting to think whether these were the sorts of readers the author expected. What do you think?

Conclusion

Readers can be just as important as writers. They're just as much a part of the history of mathematics—and even when there

---

Readers

If you're interested in the readers of a piece of historical mathematics, you can . . .

Think about who they were. What can you tell about them from the content of the mathematics, how fully explained it is, and the tone of voice the author uses?

Think about the genre of the piece. Is it a textbook, a letter, a set of private notes, etc.?

Think about whether it was meant to be read at all. Is it a piece of private mathematics?

See if you can find any clues about who actually read it. Are there any annotations? Do you know who owned it or how many copies there were, etc.?

---

weren't supposed to be any readers, you can learn something from that as well. As you did in chapter 1, you think about *how* the writer wrote—and *why*—just as much as *what, when,* or *where.*

If you can say that a particular piece of historical mathematics belongs to a particular genre, that can help you understand why it is the way it is—or help you know what to look for when reading it. The box summarizes what you've learned in this chapter about how to think about the readers of a text.

One sort of reader the author almost certainly didn't expect is us—historians of mathematics. In the final chapter we will think about what that might mean.

To think about

(1) Look again at the pictures and description of the *Young Man's Book of Knowledge.* It doesn't actually call itself a text-book—how can you tell it is one?

(2) Find a modern version of the Intermediate Value Theorem—say in a textbook or on Mathworld[3]—and a historical one (there's one in Fauvel and Gray's *Reader*). What do the differences between them tell you about the kind of readers they are for?

(3) Look again at question 4 in chapter 1. What can you tell about the readers of *The Ladies' Diary*?

[3] mathworld.wolfram.com.

(4) What do we learn about the reader(s) from this extract from a letter of Pierre de Fermat to Marin Mersenne?

Here are three propositions I have found on which I hope to erect a great building.

[Having defined the "radical numbers" as the sequence $2^n - 1$ and called n the "exponent" of such a number, Fermat asserts that:]

1 When the exponent of a radical number is compound its radical is also compound. Thus, because 6 the exponent of 63 is compound I say that 63 is also compound.

2 When the exponent is a prime number, I say that its radical reduced by unity is measured [ie divisible] by the double of the exponent. Thus because 7 the exponent of 127 is a prime number, I say that 126 is a multiple of 14.

3 When the exponent is a prime number, I say that its radical is not measured by any prime number except those which exceed by unity either a multiple of the double of the exponent or the double of the exponent. Thus, because 11, the exponent of 2047, is a prime number, I say that it cannot be measured except by a number which is greater by unity than 22, as 23, or by a number which is greater by unity than a multiple of 22; in fact 2047 is only measured by 23 or 89, from which, if you remove unity, 88 remains, a multiple of 22.

Here are three extremely beautiful propositions which I have found and proved, not without difficulty. I could call them the foundations of the invention of perfect numbers. I don't doubt that M. Frenicle got there earlier, but I have only begun and without doubt these propositions will pass as very lovely in the minds of those who have not become sufficiently hypercritical of these matters, and I would be very happy to have the opinion of M. Roberval.

—Translation from D. J. Struik's *Source Book in Mathematics, 1200–1800*, vol. 2, pp. 198–99.

(5) Who would read this? How is it different from writing for similar readers today?

We must be able to change our standard of reference at pleasure, since sometimes one is convenient, sometimes another; and Lorentz and Larmor have discussed in its essentials the transformation needed when we change from one standard, represented by distance x and time t, to another standard corresponding to distance x' and time t', which is moving relatively to the first with velocity u. These transformation equations have become tremendously significant and of universal import in the hands of Einstein,—so important that I must write them down even here, without expecting them to be as yet understood, for they lie at the bottom of the whole trouble, or (to put it otherwise) they afford nourishment to the whole tree of relativity:—

$$x' = \beta(x - ut); t' = \beta\left(t - \frac{ux}{c^2}\right); \text{ with } \beta^2(c^2 - u^2) = c^2.$$

[The writer now explains the idea of relative vesus absolute velocity.]

In order to record times and places in a way which is independent of our own position and relative movement, and which will be intelligible to people anywhere, and so to speak "true" at the moment we record them, we must allow for the time taken by light to reach us. Now that time will be $\frac{x}{c}$ if the thing observed is relatively stationary, and $\frac{x'}{c}$ or $\frac{x-ut}{c}$ if it be relatively approaching; c being the velocity of light. So calling this corrected time t', corresponding to the corrected distance x',

$$t' = \frac{x-ut}{c} = \frac{x}{c} - \frac{u}{c}t = t - \frac{ux}{c^2};$$

which after all is only the fairly obvious $\frac{t'}{t} = \frac{c-u}{c} = \frac{x'}{x}$. This gives us the true values, x' and t', for place and time of an approaching object which to the observer appears to be at x and t.

[The writer now introduces the Lorenz contraction as a physical effect resulting from the inability of the "ether" to transmit anything faster than light.]

The expressions that would otherwise be correct have, therefore, to be multiplied by a factor β, which is very nearly unity save for excessive speeds approaching the medium's critical value c. At such speeds as that, the medium's properties are becoming strained or exhausted. It cannot transmit anything with a speed greater than c; and the coefficient β rapidly approaches an infinite value as the speed c is approximated to. For all ordinary speeds, however, it is very nearly 1. We thus arrive at the equations recorded above . . . and now no one can tell whether it be the source or the observer that is moving.

The gist of the equations is that a moving observer

must take not only his distances as variable, but his times too.

—Oliver Lodge, "Einstein's Real Achievement,"

The Fortnightly Review, London, 1 September, 1921.

(6) Look again at the extract from Sophie Germain's letter to Gauss (chapter 2, question 4). What does it tell you about Gauss?

CHAPTER 5

What to Read, and Why

In 1771, Joseph Louis Lagrange published a paper on the solution of polynomial equations. He considered an equation, say $\phi(x)$, which had μ roots, x', x'', x''', etc. He considered finding a substitution $y = s(x)$ resulting in a new equation, the "reduced" equation, $\Theta(y) = 0$. And he showed that the roots of $\Theta = 0$ could then always be generated from a single function, $f(x', x'', x''', \ldots)$, by taking all possible permutations of x', x'', x''', etc.

In order for Θ to have fewer roots than ϕ (which was the point of doing the substitution), some of those permutations would have to leave f unchanged. What were the possibilities for that to happen?

> 97. Although the equation $\Theta = 0$ must be, in general, of degree $1.2.3. \ldots \mu = \pi$, which is equal to the number of permutations of which the μ roots x', x'', x''', $\ldots$ are capable, yet if it happens that the function should be such that it receives no change by one or more of these permutations, then the equation by which it is worked will necessarily be reduced to a lower degree.
>
> For let us suppose, for example, that the function $f[(x')(x'')(x''')(x^{IV}) \ldots]$ should be such that it keeps the same value on changing x' to x'', x'' to x''', and x''' to x', so that one has

$$f[(x')(x'')(x''')(x^{IV}) \ldots] = f[(x'')(x''')(x')(x^{IV}) \ldots],$$

it is clear that the equation $\Theta = 0$ will already have two equal roots; but I am going to prove that under this hypothesis all the other roots will also be equal two by two. Indeed, let us consider any other root of the same equation, which will be represented by the function

$$f[(x^{IV})(x''')(x')(x'') \ldots],$$

as this derives from the function

$$f[(x')(x'')(x''')(x^{IV}) \ldots],$$

by changing x' to x^{IV}, x'' to x''', x''' to x', x^{IV} to x'', it follows that it will also keep the same value when changing x^{IV} to x''', x''' to x' and x' to x^{IV}; so that one will also have

$$f[(x^{IV})(x''')(x')(x''') \ldots] = f[(x''')(x')(x^{IV})(x'') \ldots].$$

Then, in this case, the quantity Θ will be equal to a square θ^2, and consequently the equation $\Theta = 0$ will be reduced to this one, $\Theta = 0$, whose degree will be $\pi/2$.

One could demonstrate in the same way that, if the function

$$f[(x')(x'')(x''')(x^{IV}) \ldots]$$

is of its own nature such that it keeps the same value when making two or three or more different permutations of the roots x', x'', x''', x^{IV}, ..., the roots of the equation $\Theta = 0$ will be equal three by three, or four by four, or, etc.; and so that the quantity Θ will be equal to a cube θ^3, or to a square square θ^4, or, etc., and that consequently the equation $\Theta = 0$ will be reduced to this one, $\theta = 0$, whose degree will be equal to $\pi/3$, or equal to $\pi/4$, or, etc.

—J. L. Lagrange, "Réflexions sur la résolution algébrique des equations," in his *Œuvres*, vol. 3 (1769), pp. 370–71; my translation.

Not as advertised

That text is sometimes called the first statement of Lagrange's theorem. It's really the last sentence—"consequently the equation $\Theta = 0$ will be reduced to this one, $\theta = 0$, whose degree will be equal to $\pi/3$, or equal to $\pi/4$, or, etc."—that contains the theorem.

A modern statement of Lagrange's theorem looks entirely different:

> If H is a subgroup of a finite group G, then the order of H is a factor of the order of G.

And it's proved in a completely different way. One modern proof uses a three-step strategy: first we prove that if $x \in G$ is not a member of H, then H and xH are disjoint sets; next we prove that if $y \in G$ is a member of neither H nor xH, then xH and yH are disjoint sets. This process can be continued until G has been partitioned into one or more disjoint sets xH, yH, zH, etc. Third, we prove that each of these partitioning sets has the same number of elements as H, which implies that $g = ph$, p being the number of partitioning sets and therefore that h divides g.[1]

What are the similarities and differences, and how much do they matter?

Pause for thought

[1] David Smart, *Linear Algebra and Geometry* (Cambridge University Press, 1988), p. 254.

This is rather like the exercise we did with the passage from Newton in chapter 1.

For a start, Lagrange's statement is about the degrees of equations, not about the orders of groups and subgroups. This is certainly the most obvious difference—the two versions of Lagrange's theorem seem to be about different things.

Also, Lagrange's statement isn't presented as a theorem. He doesn't draw very much attention to it, and it comes as section 97 out of more than 100 sections. It doesn't look as if he thought it was a particularly important result. (In fact, for the purpose of solving equations, it isn't. The method—considering permutations of roots—mattered much more in the long run than this particular result.)

And Lagrange's statement isn't completely precise. Later in his paper he'd be more explicit about what generality he was claiming. But in the extract we've got, he just states that the degree of the resolvent equation may be reduced to $\pi/2$, $\pi/3$, $\pi/4$, etc. Is every divisor of π possible? Are there some that aren't available for certain values of π? He doesn't say.

Now let's look at the proof. Lagrange doesn't exactly give a rigorous proof, and he isn't really trying to. He just wants to make the result look generally believable. So he walks us through a few special cases where permutations of the roots don't affect the value of f, and shows that in those particular cases the number of distinct roots of $\Theta = 0$ is reduced by a factor equal to the degree of the permutation in question. Then he asserts that the pattern can be generalized. (It doesn't help that he makes trivial errors when considering his special cases—did you spot any?)

Finally, is Lagrange's statement precisely equivalent to the modern one? It's quite hard to say, precisely because Lagrange's statement just isn't about groups. You might well feel that it is somewhat less general than its modern cousin.[2]

You might be wondering whether there's any sense in calling our passage "the first statement of Lagrange's theorem." Yes, it does contain something equivalent to the idea that the order of a subgroup divides the order of its parent group—an idea that was put into those terms rather later than this passage. But on the other hand, there are all sorts of differences—what the passage is about, how precise the statement is, whether it's proved properly or called a theorem, and even whether it's really equivalent to its modern "equivalent." Personally, I think this does deserve to be called a statement of Lagrange's theorem, just. But you could certainly argue it either way.

Quite often in the history of mathematics you'll be faced with texts that don't quite perform as advertised. Either the context is different (equations rather than groups, say), the content is different (how general is the result?), or both. The box sum-

---

[2] There's more on this, if you're interested, in Richard L. Roth's article, "A History of Lagrange's Theorem on Groups," in *Mathematics Magazine* 74 (April 2001), pp. 99–108. Lagrange's statement is essentially this, in modern notation:

> If the function Θ of μ variables is acted on by all $\mu!$ possible permutations of the variables, these permuted functions will take on a number of distinct values, r, say. r may be smaller than $\mu!$, but it must be a divisor of $\mu!$

And in terms of groups it comes to this:

> Let S be the set of functions involving n variables formed by all permutations of the n variables: let S_n be the group acting on S, whose group action is that arising from permuting the variables in these functions (thus, S_n is the symmetric group of size n). Then the size of any orbit in S_n is a divisor of $n!$.

Not as Advertised

If a piece of historical mathematics doesn't look the way you'd expect, you can ask . . .

What's the context? What sort of mathematical objects (numbers, equations, graphs, groups, etc.) is it about?

Is it a theorem or a casual remark, or something in between?

Is it proved or argued for, or just stated?

Is it precise?

How general is it? Is it really equivalent to its "modern equivalent"?

What does the author use it for? Why is the author interested?

---

marizes some useful questions to ask to help think about what's going on in situations like that.

Mathematical change

Where does all this get us? Certainly, some mathematical firsts are not quite what they seem—they involve a bit of hindsight, a bit of "reading into" the original text something that is only just there.

We can say more than that, though. In a way this kind of thing is what the history of mathematics is about—the fact that mathematics changes over time, and we can learn to recognize the ways it changes and study them and understand them. We've just seen some of the questions you can ask. Let's put them in more general terms.

Has the context changed? The essentials of Lagrange's theorem can be expressed in terms of equations or in terms of groups. The same is true of any mathematical idea: there are different contexts in which it can appear, and it might look very different while still being essentially the same. Sometimes it first appears in one form but becomes well known in another.

That's one common way mathematics can change—and you can always translate a historical version into modern terms to check that it's essentially the same as its modern version, and spot the differences to see whether it's *exactly* the same.

Has the idea become more important, or less? When you change the context, an old idea might become trivial—or, like Lagrange's theorem, it might become more important. And, over time, results that were once new and exciting become the stuff of basic textbooks or even fall out of view altogether. You can look at the way the original author presents the idea to find out how important it was for that author—and then see if you can find a version in a modern textbook. If you can, is it given more attention or less? If you can't, what does that tell you?

Has the idea become more precise? You'll very often find statements in historical mathematics which seem vague by modern standards. Remember our example from Newton in chapter 1, where some of the terms seemed quite unclear and it was possible to find a counterexample to the result. Often new mathematics comes from revisiting existing results and proofs and finding gaps to fill in. In the eighteenth century mathematicians wanted to understand how the calculus could be made to work rigorously, for example. And in the late nineteenth and early twentieth centuries, some mathematicians tried to formulate the foundations of mathematics in a more rigorous way than had ever been done before.

So, an increase in rigor or precision is another of the main ways mathematics changes. And you can always check through a piece of historical mathematics to see whether it's as rigorous as a modern version would be—and think about how and why it isn't.

Is it proved differently? There are two different things that can happen. First, in older mathematics you often see statements justified by a vague appeal to "symmetry," or with a suggestion that a particular pattern continues, or—as in Lagrange's theorem—a quick look at some special cases. Nowadays the same result would probably get a rigorous proof. That's like the increase in rigor and precision we just talked about, and you can study it the same way—by spotting differences and thinking about how and why they've come about.

Second, you might find a whole different strategy of proof compared with what you're expecting. Even today, some results have more than one different strategy of proof in the textbooks. In this case, too, it can be interesting to think about why the older proof is different. Is it to do with a different mathematical context (that's the case for our example from Lagrange), or is it because the original writer didn't have some of the mathematical tools the modern proof uses?

Finally, *has the idea become more general?* This is another very common way that mathematics can change, and we saw it with Lagrange's theorem. The theorem now covers all finite groups, but Lagrange's own version of it was not as general as that. Sometimes, of course, a theorem is renamed when it becomes much more general (can you think of an example?) When you're looking at a piece of historical mathematics, you can always ask whether it covers all the cases you would expect it to—and if not, why that might be.

So there are all sorts of ways mathematics can change. Things get more *precise*, more *general*, more *rigorous*. *Strategies of proof* change, and sometimes ideas are transplanted from one mathematical *context* to another. On top of all that there are the actual new pieces of mathematics—things like calculus, or groups. They don't always come from absolutely nowhere, but they change mathematics—suddenly things are being done in mathematics that just weren't being done a few years earlier.

Mathematical Change

When you're thinking about how a mathematical idea has changed, or how a piece of historical mathematics shows you those changes, you can ask . . .

Has the context changed?

Has the idea become more important, or less?

Has the idea become more precise?

Is it proved differently—more rigorously, or just plain differently?

Has it become more general?

(And the example from Lagrange shows how an individual mathematical idea can survive across a change like that.)

These kinds of changes are often messy and always unpredictable. But that's what makes the history of mathematics worth studying—more so than if it just consisted of a neat sequence of new discoveries with names and dates attached to them. As it is, there's always more to find out about the mathematical past, and always new questions to ask.

Significance

Why are we reading this example from Lagrange? Why are we reading any of the examples in this book, or anywhere else?

One answer is because they're significant. What do we mean by that?

Look again at the example from Lagrange. Why do you think it might be called historically significant?

Pause for thought

I can think of three reasons, although you might have come up with some more. One is that, despite all the differences we just talked about, this passage contains the first statement of Lagrange's theorem, or something very like it. Mathematical firsts can be a little bit misleading, and there's a lot more to the history of mathematics than them—but they are important all the same. If a piece of writing contains a genuine new mathematical idea, it probably deserves to be called significant.

Another reason is that it takes a new approach to its subject. This wasn't something we discussed above, but if you found out some background about Lagrange, or the theory of equations, or if you found this extract in another textbook, you'll have learned that Lagrange's idea of studying the permutations of the roots of an equation in this way was basically new. Not only that, but this approach led to the proof, a few decades later, that the quintic equation was in fact unsolvable in general. And it led eventually—and more indirectly—to group theory and Galois theory. You saw in chapter 2 a little of what had been done in that line 60 years later.

New approaches can be dead ends, of course—maybe most of them are. But a new approach that actually produces results—solves problems, leads to more new mathematics, or in this case a whole new branch of mathematics—surely deserves to be called significant.

A third reason—and it's quite similar to the second—is that this piece of writing was influential. If Lagrange had written this piece of mathematics—with its first statement of Lagrange's theorem and its new approach to solving equations—but then locked it away and not published it, what would have happened? Not much. The consequences we just talked about only happened because people read this piece of writing and took it seriously. It changed what mathematics they did and how they did it. It was influential, in other words. Influential mathematics nearly always deserves to be called significant, I think.

So significance can be about firsts, about new approaches, and about being influential. Not all significant pieces of mathematics have all three qualities. You can probably think of examples.

A textbook, like the one by l'Hôpital that we looked at in chapter 4, might be very influential—it might change the way a whole generation of mathematicians looks at and thinks about a particular branch of mathematics—without saying anything actually new. Or a first might be significant for historians even if it wasn't read or wasn't understood at the time—significance can sometimes be a matter of hindsight.

But not all historical mathematics *is* significant. And perhaps there's a second kind of significance, where something can be historically significant without being mathematically

---

Significance

To decide whether a piece of mathematics is significant, you can ask . . .

Is it a first? Does it state some result for the first time? (Or ask a new question, define new notation, describe a new mathematical object, etc.?)

Does it take a new approach to a problem? Did it lead to more new mathematics, or to solutions of long-standing problems?

Was it influential? Did it affect what mathematics other mathematicians did? Or how a particular branch of mathematics was thought about?

And remember: not all of it *is* significant. Are we reading it for reasons that go beyond "significant mathematics"?

---

significant. Some historians (I'm one of them) delight in investigating mathematical writing that contains little or no important or novel mathematics: popular textbooks, self-instruction manuals, and schoolbooks like the one we looked at in chapter 3, or old almanacs and popular magazines with mathematical news or puzzles in them. These kinds of writing tell us a lot about what sort of mathematics ordinary people knew about, learned, or used; they are certainly significant for a historian who wants to know about popular experiences of mathematics. But they're not significant in the sense of containing significant mathematics. In this book I've included a few examples of these kinds of things, to show that there's a lot more to the history of mathematics than the obvious things that are often talked about—and some of those nonobvious things can be both informative and tremendous fun.

Historians

And then, why are *you* reading this particular piece from Lagrange? Who asked you to? Who decides that something is significant, or interesting?

Well, I do. Often it's the people who write textbooks—historians of mathematics, in other words—who choose what gets read and what doesn't. Historians of mathematics can do a lot of things to a piece of historical mathematics. You've seen how they can translate it, edit it, maybe tamper with the notation. They can use the benefit of hindsight to say that it's significant (or not). You've now learned to recognize when those things are going on.

But maybe the main thing historians do is the most obvious one—they choose which pieces of historical mathematics to study at all. You can't look at everything—you've got to choose. One choice is about which pieces of mathematics to read: Lagrange or Euler? Group theory or probability? However you choose, a lot of mathematical writing and a lot of mathematicians simply drop out of the picture.

A bigger choice is which cultures to look at. Mathematics developed, sometimes at different times and sometimes at the same time, in many different cultures. In this book I've focused on just one—Europe between about 1550 and about 1900. But you could choose differently—or you could compare two or more cultures. In any case, you're making a decision about what goes into your history of mathematics and what stays out of it.

Even when those two decisions are made, you still have to make more detailed decisions. Which bit of the *Principia* is it interesting to look at? Which edition do you choose? Which translation? These kinds of choices can make a real difference in how the *Principia* looks—and so in how the history of mathematics looks as a whole.

And, finally, once you've chosen what to read, you have to decide what to do with it. Our example from chapter 2, Galois' last letter, gets used in a lot of different ways. It features in the plot of several plays and novels. Some books use it to show something of Galois' work. Some use it to illustrate his character. I used it to show how you can learn about an author and his context from a text. Any of those uses is perfectly reasonable—but they all start with the same text and make a different sort of history of mathematics out of it.

So historians—and that means you, too—can change things, both by *what* they read and *how* they read it and use it. This is another thing that makes studying the past fascinating and worthwhile. We can never totally disentangle ourselves—our own choices and our own ideas—from our study of the past.

What can you do about this? I've suggested before that if you suspect a text has been tampered with, you can find a different version of it and check. You can also always ask yourself the questions I've just mentioned. Who chose this text? This culture? This particular passage from that text? It might be the writer of a textbook or a teacher—or it might be you yourself. Why those choices, rather than any others? What are we doing with the texts we've chosen, and why? What sort of history of mathematics are we putting together?

Historians

If you're asked to read a particular piece of historical mathematics you can ask . . .

Why this culture?

Why this author, this book, or this part of mathematics?

Why this particular extract?

What are we doing with this text? What sort of history of mathematics—a novel, a history of group theory, a study of an individual person—are we doing?

And you can think about how those choices might be made differently.

These can be interesting things to think about. There are always different ways of making these choices—so there are always new ways of doing the history of mathematics. And thinking about those different choices can be a way to start exploring historical mathematics on your own.

Conclusion: *What to read, and why*

There's a box below which summarizes the questions you've learned to ask in this chapter, and there's another one right at the end which summarizes this whole book. You deserve congratulations—you've learned a lot of different ways to read

---

What to Read, and Why

When you're presented with a piece of historical mathematics, you can ask . . .

Does it perform as advertised? If not, what is different from what you'd expect—and why does it matter?

How have mathematical ideas changed? Is it in their context, their importance, their precision, or their generality—or something else?

Is it significant? Why? Is it a first, a new idea, or an influential piece of writing—or not?

Why are you reading this—this culture, this author, this book, this extract? What are you doing with it? And how might you choose differently?

---

a piece of historical mathematics. That means you've learned a lot about how to be a historian of mathematics. You can now feel confident about tackling many different kinds of mathematical writing and making some sense of them—thinking, *what on earth was that?* and being able to come up with some answers.

I hope you'll find the various boxes along the way, with suggestions of what to do in particular situations, useful to refer

---

How to Read

When you're reading a piece of historical mathematics, you can ask . . .

What does it say? Can I put it into modern terms? And if I do, what surprises come up?

Who wrote it? Where, when, and why? What can the library and the web tell me about the author and the context?

What physical object does this text come in? Who has left a mark on it apart from the author? Is it a prestigious object, widely disseminated, and durable—or not?

Who was meant to read it? How can you tell? What genre of writing is it? Or is it a private piece of writing?

Why are you being asked to read it? What mathematical changes does it illustrate? What choices have you or your teacher made here?

---

back to. As you've seen, there isn't one right way to read, but now you have a toolkit of ideas that will help—and that will help you to ask your own questions when you read. We've looked at short extracts in this book, mostly with fairly straightforward mathematics—but I hope what you've learned will also help you if you want to read a whole book, or something more technical. (You don't have to completely understand the technical details in order to ask a lot of the questions we've come up with.)

Most of all, I hope you continue to read historical mathematics, and to enjoy it.

To think about

(1) Find passage 18.C2 in Fauvel and Gray's *Reader*. It's called "Cantor's definition of the real numbers." Does it *really* contain a definition of the real numbers? If so, what exactly is it? What are the reasons for this passage being included in the *Reader*?

(2) Here are two versions of a passage from Newton's *Principia*.

> For if the action of an agent is reckoned by its force and velocity jointly, and if, similarly, the reaction of a resistant is reckoned jointly by the velocities of its individual parts and the forces of resistance arising from their friction, cohesion, weight, and acceleration, the action and reaction will always be equal to each other in all examples of using devices or machines.
> —Translated by Cohen and Whitman, 1999.

Work done on any system of bodies has its equivalent in the form of work done against friction, molecular forces or gravity, if there be no acceleration; but if there be acceleration, part of the work is expended in overcoming resistance to acceleration, and the additional kinetic energy developed is equivalent to the work so spent.

—Translated by P. G. Tait, 1877.

(Both are quoted in Cohen and Whitman, *The Principia*, p. 119.)

Why are they so different? Think about the image of Newton and the *Principia* that they give, and about the fact that Tait (what can you find out about him?) was not a historian, but Cohen and Whitman were.

(3) Look again at my examples in chapter 3. Pictures of a minor eighteenth-century textbook are not normal things to find in a book about the history of mathematics. What am I up to? What ideas am I trying to get across (apart from the ones I state in the accompanying text)?

(4) Take a look at *Ramanujan's Notebooks*—the published version, edited by Bruce C. Berndt (most of the volumes are on Google Books). Choose a particular entry and discuss what the editor has done, and how and why the presentation here is different from how it would have been in the original notebook. What choices has the editor had to make, and what sort of history of mathematics has his choices produced?

(5) Find out as much as you can about how—and why— Lagrange's name came to be attached to Lagrange's theorem.

BIBLIOGRAPHY

This bibliography lists some of the general works I used when writing this book or have referred to during the course of it; there are also some chapter-by-chapter suggestions for further reading on the topics discussed, some of them quite specialized, some more general or taking the ideas much further afield.

Primary Sources

Early English Books Online: eebo.chadwyck.com.

Eighteenth-Century Collections Online: galenet.galegroup.com/servlet/ECCO.

John Fauvel and Jeremy Gray, eds., *The History of Mathematics: A Reader* (Palgrave Macmillan, Basingstoke, U.K., 1987).

Gallica: gallica2.bnf.fr.

Victor J. Katz, ed., *The Mathematics of Egypt, Mesopotamia, China, India, and Islam* (Princeton University Press, Princeton, N.J., 2007).

Jacqueline Stedall, *Mathematics Emerging: A Sourcebook 1540–1900* (Oxford University Press, Oxford, 2008).

D. J. Struik, ed., *A Source Book in Mathematics, 1200–1800* (Harvard University Press, Cambridge, Mass., 1969).

Textbooks

Victor J. Katz, *A History of Mathematics: An Introduction* (Addison-Wesley, 3rd edition, Reading, Mass., 2009).

William Dunham, *The Calculus Gallery: Masterpieces from Newton to Lebesgue* (Princeton University Press, Princeton, N.J., 2004).

Ivor Grattan-Guinness, *Companion Encyclopedia of the History and Philosophy of the Mathematical Sciences* (2 vols., Routledge, London, 1993).

Sources of Journal Articles

JSTOR: www.jstor.org.
Scopus: www.scopus.com.
Web of Science: apps.isiknowledge.com.

Chapter 1. What Does It Say?

Isaac Newton, trans. I. Bernard Cohen and Anne Whitman, *The Principia: Mathematical Principles of Natural Philosophy* (University of California Press, Berkeley, 1999).
The Newton Project: www.newtonproject.sussex.ac.uk.
D. T. Whiteside, ed., *The Mathematical Papers of Isaac Newton* (Cambridge University Press, Cambridge, 1967–1981).

Chapter 2. How Was It Written?

Charles Coulston Gillispie (ed.), *Dictionary of Scientific Biography* (16 vols., Scribner, New York, 1970–1980).
H. C. G. Matthew and Brian Harrison (eds.), *Oxford Dictionary of National Biography: From the Earliest Times to the Year 2000* (60 vols., Oxford University Press, Oxford, 2004): www.oxforddnb.com.
Laura Toti Rigatelli, trans. John Denton, *Evariste Galois 1811–1832* (Birkhäuser, Basel, 1996).
Jenny L. Presnell, *The Information-Literate Historian: A Guide to Research for History Students* (Oxford University Press, Oxford, 2007).

Chapter 3. Paper and Ink

A timeline of writing and printing in history: www.xs4all.nl/~knops/timetab.html.
Florian Cajori, *A History of Mathematical Notations* (2 vols., third edition, Myers Press, Warrington, U.K., 2007).
COPAC: copac.ac.uk.
David Finkelstein and Alistair McCleery, *An Introduction to Book History* (Routledge, London, 2005).

Adrian Johns, *The Nature of the Book: Print and Knowledge in the Making* (University of Chicago Press, London, 1998).

Latin Place Names: net.lib.byu.edu/~catalog/people/rlm/latin/names.html.

Library of Congress Catalog: catalog.loc.gov.

Michael Twyman, *The British Library Guide to Printing: History and Techniques* (British Library, London, 1998).

Chapter 4. Readers

Alberto Manguel, *A History of Reading* (Viking, London, 1996).

Guglielmo Cavallo and Roger Chartier (eds.), *A History of Reading in the West* (Polity Press, Cambridge, 1999, 2003).

Chapter 5. What to Read, and Why

Ivor Grattan-Guinness, *Landmark Writings in Western Mathematics 1640–1940* (Elsevier, Amsterdam, 2005).

J. D. Jackson, Examples of the zeroth theorem of the history of science, *American Journal of Physics* 76 (2008), 704–19.

Harold Bloom, *How to Read and Why* (Fourth Estate, London, 2000).

Italo Calvino, *Why Read the Classics?* (Pantheon, New York, 1999).

INDEX